BURLEIGH DODDS SCIENCE: INSTANT INSIGHTS

NUMBER 100

Ensuring the welfare of broilers

I0130680

burleigh dodds
SCIENCE PUBLISHING

Published by Burleigh Dodds Science Publishing Limited
82 High Street, Sawston, Cambridge CB22 3HJ, UK
www.bdspublishing.com

Burleigh Dodds Science Publishing, 1518 Walnut Street, Suite 900, Philadelphia, PA 19102-3406, USA

First published 2023 by Burleigh Dodds Science Publishing Limited
© Burleigh Dodds Science Publishing, 2024, except the following: Chapter 3 remains the copyright of the author. All rights reserved.

British Library Cataloguing in Publication Data
A catalogue record for this book is available from the British Library

ISBN 978-1-80146-673-8 (Print)
ISBN 978-1-80146-674-5 (ePub)

DOI: 10.19103/9781801466745

Typeset by Deanta Global Publishing Services, Dublin, Ireland

Contents

Series list

Chapter 1

Ensuring the welfare of broilers: an overview

T. B. Rodenburg, Wageningen University, The Netherlands

1   Introduction

Broilers have been intensively selected for high feed efficiency and fast growth over the past 60 to 70 years. This selection has been extremely successful, increasing broiler growth rate between 1957 and 2005 by over 400%, with a concurrent 50% reduction in feed conversion ratio (Zuidhof et al., 2014). However, this selection for highly efficient, fast-growing broilers had negative effects on broiler welfare. Activity levels of fast-growing broilers are low, especially during the second half of their life. Leg health can be problematic, with issues such as lameness, and high incidences of footpad dermatitis and hock burns playing a major role. In the past, also tibial dyschondroplasia caused many health and welfare problems, but the incidence of this metabolic disease has been strongly reduced by selective breeding. Fast-growing broilers are also more sensitive to heart and circulation problems compared with slower-growing genotypes, especially when placed in a suboptimal environment. When subjected to a mild cold stress, the incidence of mortality due to heart and circulation problems in fast-growing broilers was 15% compared to 2% in slower-growing genotypes (van Horne et al., 2004). The majority of mortality cases was caused by ascites, a disease causing circulation problems resulting

http://dx.doi.org/10.19103/AS.2016.0011.25

in fluid accumulation in the peritoneal cavity. Regarding reduction of ascites incidence in commercial broilers, substantial progress has been made in recent years (Thiruvenkadan et al., 2011).

In 2000, the EU Scientific Commission on Animal Health and Welfare (SCAHAW) published an authoritative report on the welfare of broilers, which was revised in 2012 (de Jong et al., 2012). The main recommendation of the 2000 version of the report was that, *'Breeders should give a considerably higher priority to health variables in the breeding index, if necessary at the expense of the selection pressure for growth and feed conversion. Breeding which causes very poor welfare should not be permitted and breeders should be responsible for demonstrating that the standards of welfare in the chickens produced by them are acceptable'* (SCAHAW, 2000: 110). In the European Union this has led to the development of systems where slightly slower-growing broiler hybrids are used, which show less welfare problems. In The Netherlands, for instance, this has resulted in the development of a new production system with a slower-growing genotype, more space per bird and access to a covered veranda; products from this system are sold at a premium price under a welfare label supported by the Dutch organisation for animal protection. Apart from genetics, also the relatively high stocking densities in broiler production (over 15 birds/m²) have received a lot of attention in welfare research. Especially in systems with poor management or housing conditions, high stocking densities can lead to poor litter quality and welfare problems, such as poor leg health and increased mortality rates. The relationship between high stocking density and poor welfare was the reason for the EU to limit the stocking density in broilers to 33 kg/m². Only if the producer could show that the company could manage broilers at a higher stocking density without high mortality for seven subsequent flocks (below 3.5% for a slaughter age of 42 days), the producer could be allowed to keep broilers at a higher stocking density to a maximum of 39 kg/m². This focus on animal-based measures, such as leg health and mortality, is also supported by newly developed methods to assess broiler welfare, such as the European Welfare Quality method (Welfare Quality Assessment Protocol for Poultry, 2009). Furthermore, practical methods to assess broiler welfare in commercial flocks, such as the transect method (Marchewka et al., 2013), have further increased the opportunities for welfare assessment. Together, increasing attention for broiler health and welfare in breeding and regulation of stocking density in the European Union have led and are expected to lead to an improvement in broiler welfare, while still allowing efficient broiler meat production.

2 Genetic selection

As indicated in the introduction, genetic selection for high feed efficiency and fast growth has been extremely successful in broilers. However, this selection for highly efficient, fast-growing broilers had negative effects on broiler welfare (Rauw et al., 1998; Julian, 1998; Sandoe et al., 1999). In 2000, this led to the SCAHAW report concluding that breeders should put more priority on broiler health, if necessary at the expense of growth and feed conversion and that breeding that causes poor welfare should not be permitted. So, what happened since that recommendation in 2000? One approach has been for breeders to focus more on reducing health and welfare issues. The main focus here has been on improving leg health and reducing lameness. Kapell et al. (2012b) describe that despite low heritabilities for traits such as tibial dyschondroplasia and hock burns and negative correlations with efficiency, substantial progress has been made over the past 25 years in improving leg health through balanced breeding programmes. Similarly, Rekaya et al. (2013) concluded that it is feasible to improve leg health without large negative effects on growth performance. Also regarding footpad and hock dermatitis, reduction through genetic selection seems feasible (Ask, 2010). Ask indicates that contact dermatitis would likely increase when genetic selection for improved growth rate would continue, without attention for leg health in the breeding goal. To be able to improve leg health, it is of course important to record leg health traits not only in the breeding stock, but also in commercial broiler flocks (Hocking, 2014). Kapell et al. (2012a) compared genetic selection for improved footpad health in a pedigree breeding environment and in a sib-testing environment, more similar to a commercial broiler operation. They found a strong genetic correlation of around 0.80 between footpad dermatitis measured in both environments. This indicates that it should be feasible to improve footpad health in commercial flocks through selective breeding in the pedigree birds, although they are kept in a more optimal environment than commercial flocks. Similar approaches as for improving leg health have been taken for reducing ascites in broilers. Pakdel et al. (2005) showed that by including reduction of ascites into the breeding goal, the improvement in body weight gain was reduced from 130 to 111 g, but it was possible to reduce the incidence of ascites in response to cold stress using this second breeding scheme. These studies show that there is potential to improve broiler health and welfare through genetic selection, without strong negative effects on growth performance. At the same time, the clear relationship between genetic selection for fast growth and broiler health and welfare has led to societal pressure to move to broiler production systems using slower-growing genotypes (Dawkins and Layton, 2012), especially in the European Union. For instance, the study by Knowles et al. (2008) indicated that in a comprehensive survey of broiler flocks in the United Kingdom almost 28% of the birds showed poor locomotion at 40 days of age and 3% was almost unable to walk. Bird genotype was identified as one of the main risk factors for poor locomotion. Slower-growing genotypes, such as those used in organic production, are known to have much lower incidences of health and welfare problems (Bessei, 2006). For organic broiler meat production, EU regulations stipulate that slower-growing genotypes should be used with a growing period of 12 weeks.

This makes organic broiler meat production relatively costly due to the high feed costs. However, the lines used to breed the slower-growing strain can also be used to breed a hybrid that shows intermediate growth (6 to 8 weeks to slaughter weight) and still shows a much lower incidence of health and welfare issues compared with fast-growing hybrids (van Horne et al., 2004). This intermediate type is now being used in European broiler production that aims at a market between conventional and organic. This type of broiler meat is usually sold within a specific welfare assurance scheme, such as the Freedom Foods Standards in the United Kingdom, or the 'Beter Leven' standards in The Netherlands. Within these schemes, systems receive a specific number of stars depending on their attention to welfare issues. Within the Dutch middle-market system with one star, producers are required to use the intermediate growing broilers. Recently, Gocsik et al. (2016) compared Dutch conventional, middle-market and organic systems and found that the middle-market system received the highest Welfare Quality score, indicating the best welfare, compared with the other systems (Welfare Quality Assessment Protocol for Poultry, 2009). Regarding cost-efficiency, these middle-market systems outperformed free-range and organic farms, so these systems may be an attractive alternative for efficient broiler meat production with good health and welfare (Gocsik et al., 2016).

3 Stocking density

As indicated in the introduction, a second major topic regarding broiler welfare is the stocking density at which the birds are kept, usually expressed in kilograms of live weight per metre squared. Broilers are kept at relatively high stocking densities of over 15 birds/m^2 or 30 kg/m^2. Towards the end of the growing period, when birds approach slaughter weight and activity levels are low, these high stocking densities can lead to welfare problems. Especially in systems with poor management or housing conditions, high stocking densities can lead to poor litter quality and welfare problems, such as poor leg health and increased mortality rates. However, an influential study by Dawkins et al. (2004) showed that broiler welfare is in fact influenced more by housing conditions and farm management than by stocking density per se. They studied broiler welfare at ten major broiler-producing companies in the United Kingdom. For the study, each producer managed flocks at five different stocking densities, ranging from 30 to 46 kg/m^2. Although there were clear differences in welfare indicators across flocks and producers, these could not be linked to stocking density. Factors such as litter moisture and air ammonia had strong effects on the welfare indicators measured, but some producers also managed to maintain a good litter quality and broiler welfare at the higher stocking densities. A critique on the study by Dawkins et al. (2004) has been that the low stocking density of 30 kg/m^2 was still too high to see any major effects, as earlier studies including stocking densities lower than 30 kg/m^2 did find clear effects on behaviour and welfare (Bessei, 2006). Having said that, the study by Dawkins et al. (2004) does underline the importance of good management: in a modern, well-ventilated and well-heated house with good management it is possible to keep broilers at a higher stocking density than in more basic housing systems. In the European Union, this has led to a rather unique Council Directive (2007/43/EC), where for the first time in history animal-based welfare measurements are included in the legislation. In that Directive, stocking density in broilers is limited to 33 kg/m^2. Only if the producer can show that the company can manage broilers at a higher stocking density without high mortality for seven subsequent flocks (mortality below 3.5% for a slaughter age of 42 days), the producer can be allowed to keep broilers at a higher stocking density of maximum 39 kg/m^2. The legislation also includes specifications about temperature, ammonia and humidity levels inside the house and includes a welfare inspection at slaughter, focusing on signs of footpad dermatitis, parasitism and systemic illness at the farm of origin. This seems a promising direction for legislation on animal welfare: by focusing on animal-based parameters at the farm or at slaughter, more precise information regarding the actual welfare status of the animals can be obtained, compared with just focusing on stocking density or other more descriptive parameters.

4 Case study: hatching environment and early feeding

Currently, major changes are taking place regarding hatching environment and early feeding. Traditionally, broiler chicks hatched at the hatchery remained in the hatcher without access to feed and water until all chicks had hatched. Then they would be processed and transported to the broiler farm, where they would be placed and would receive feed and water. The relatively long period without feed and water (up to three days for early hatching chicks) could have negative effects on growth performance, but is also criticised by animal protection organisations as being inhumane. In recent years, this has led to the development of systems that do not have this long fasting period between hatching and placement, such as the Patio system from Vencomatic (Eersel, The

Figure 1 Broilers hatching in the Patio system (left panel) and the X-Treck system (right panel). In both systems, the eggs trays are suspended over the litter (photos courtesy of Vencomatic Group, Eersel, The Netherlands).

Netherlands) and the HatchBrood system from HatchTech (Veenendaal, The Netherlands). HatchBrood is a system where broiler chicks receive access to feed and water in a lighted environment directly after hatching. Lamot et al. (2014) showed that early access to feed led to a higher feed intake and body weight gain during the first week of life, but the effect disappeared thereafter. In the Patio system, eggs that have been incubated for 18 days are placed on trays at 50 cm above the litter floor. When chicks hatch, they drop down into the litter and once they dry up, they have immediate access to feed, water and conspecifics (Fig. 1, left panel). They do not have to be transported as young chicks and can remain in the Patio system until processing, or they can be moved to a finishing house during a less sensitive phase of the growing period. First results indicated a similar hatchability as in standard hatchers and low mortality rates (between 1% and 2%) (van de Ven et al., 2009). Van de Ven et al. (2013) showed positive effects of early access to feed on chick development and physiology. An added advantage of this type of systems seems to be the reduced need for antibiotic treatment, thereby limiting the risk of increased antibiotic resistance. The causes for this reduced need of antibiotics are still being investigated, but could be related to early access to feed and earlier establishment of gut microbiota and to the reduced stress after hatching. As the Patio system requires a major investment in housing equipment, alternatives have been developed that also allow broiler chicks to hatch at the farm in standard broiler houses, such as the X-Treck system (Vencomatic, Eersel, The Netherlands). In the X-Treck system, a frame is suspended above the litter and the trays with hatching eggs are placed on this frame. After hatching, the frame can be winched up and the system can be managed like a standard broiler house (Fig. 1, right panel). This type of systems is developing rapidly and is increasingly used in commercial practice. Performance, health and welfare of flocks in these systems is currently being monitored and compared to flocks from standard hatcheries. If these systems can indeed lead to better health and welfare and lower antibiotic use, this will mean a major step forward in broiler welfare. Of course, hatching at the broiler farm also has disadvantages, such as the inability to spray vaccinate upon hatching and the impossibility to segregate males and females. These disadvantages should be taken into account when further developing these systems. Furthermore, this type of system requires better management skills of the broiler farmer, especially in the period around hatching. De Jong et al. (2015) recently hatched broiler chicks from the same parent stock flock at different hatcheries and showed that these chicks had a very different performance depending on the hatchery at which the chicks hatched. The same is probably true, and likely even more pronounced, for commercial broiler farms.

5   Case study: alternative and free range systems

Apart from conventional broiler production systems, several alternative and free-range systems for broiler production exist. For instance, under the Dutch 'Beter Leven' standards, described earlier in the section on genetics, systems that receive one star have to provide their birds with access to a covered veranda. Ruis et al. (2004) found that these verandas are well-used, with about 13% of the flock present in the veranda on average, and they were mainly used for active behaviours, such as foraging and dust bathing, while indoor levels of resting increased. This same pattern of active behaviour outside and passive behaviour inside the house was observed by Fanatico et al. (2016) in a study focusing on free-range broilers. Outdoor access in free-range systems offers many extra opportunities to the birds that may be difficult to offer in an indoor system: opportunities for foraging on fresh plant and animal material, a large space allowance (4 m² per bird according to European legislation), fresh air and daylight. It can be questioned whether the space allowance of 4 m² per bird is meaningful for the birds, as this seems very large compared with space inside the broiler house. Furthermore, broilers will often stay close to the house and use only a part of the outdoor run, especially when limited cover is provided in the outdoor run. Outdoor access also presents a risk for broiler health and may result in high mortality rates due to predation. Health risks mainly consist of contracting infectious diseases, such as avian influenza, due to increased contact with wild birds or their droppings in the outdoor run, but can also consist of parasitic infections (for laying hens, see Jansson et al., 2010), food-borne pathogens such as *Campylobacter* (Rodenburg et al., 2004) or enteric infections (Van Overbeke et al., 2006). This makes that free-range and organic systems are generally more popular in areas of Europe that are less densely populated and with relatively low concentrations of poultry production. Regarding mortality due to predators, foxes can be kept out by proper fencing. Mortality due to birds of prey is more difficult to control; however, providing shelter seems to play an important role. Dal Bosco et al. (2014) showed that

Figure 2 Slow-growing broilers using the short-rotation coppice of willows for shelter in the outdoor run (photo courtesy of Lisanne Stadig, ILVO, Melle, Belgium).

by providing free-range broilers with high vegetation, almost no mortality due to predation was recorded (0.4%), while on average 2.6% mortality due to predation was recorded in a barren free range. Providing cover in the free range also seems to be key to attract a large percentage of the flock to use the free range (Dawkins et al., 2003). Furthermore, the type of cover seems to matter, for example, artificial or natural cover. Stadig et al. (2016) compared the use of the outdoor run between broilers with artificial, wooden shelters and broilers with access to short-rotation coppice with willows (Fig. 2). Broilers in the short-rotation coppice had a higher use of the outdoor run than broilers with artificial shelters (43% vs 35% of the flock). Percentages were relatively high compared with other studies, which may be explained by the fact that broilers were housed in small, mobile houses placed in the free range. Fanatico et al. (2016) reported a much lower average percentage of range use of 12.9% in broilers kept in more conventional floor housing. They compared range use in broilers with and without structural enrichments and showed that they were able to draw birds further away from the house by offering enrichments (13 vs 6% of birds venturing 15–30 m from the house).

6   Summary: how research can contribute to enhanced and sustainable broiler production

Both case studies illustrate how research can contribute to sustainable broiler production. The various systems that allow the chicks access to feed and water directly after hatching are advantageous because chicks that hatch early do not have to wait in the hatcher without access to feed and water until all chicks have hatched. The long fasting period in conventional hatcheries is increasingly criticised by animal protection organisations in Europe. Furthermore, early access to feed seems to benefit early muscle development (Powell et al., 2016a), and to limit development of adipose tissue (Powell et al., 2016b) in comparison with 24-hour feed restriction. Early access to feed will also favour an early colonisation of the gut, which may favour gut health and immunity. This may also be the reason why broilers from this type of systems seem to require fewer antibiotic treatments. For the system where the broilers hatch on the farm, an added advantage is that chicks are not subjected to the stress of processing at the hatchery or transporting to the farm. Although long-term, negative effects of transport of broiler chicks seem limited (Jacobs et al., 2016), the situation where handling and transport are avoided seems much more natural and could have positive effects on welfare by reducing stress. As indicated previously, of course this type of system where the chicks hatch at the broiler farm requires more from the management skills of the farmer. Temperature of the house and floor becomes more critical if a producer chooses for on-farm hatching compared to placement of day-old chicks. Broiler farmers need to be trained in managing the hatching process and the early life phase. However, first results of this type of broiler production system are promising and the system seems to benefit broiler health, welfare and productivity as well as societal perception. The second case study focused on alternative and free-range broiler production system. As indicated previously, this type of middle-market systems is gaining importance in the European Union. Systems that offer a covered veranda or free-range access are perceived as more welfare-friendly than conventional broiler production systems and are supported by animal protection organisations. One of the major benefits of these systems is that slower-growing genotypes are used compared to the fast-growing broiler used in conventional broiler production. This seems to result in fewer health and welfare problems and lower mortality rates in these alternative systems. Interestingly, these middle-market systems seem to outperform conventional and organic systems not only regarding their welfare score, but also regarding cost-efficiency (Gocsik et al., 2016). In that way, these middle-market systems may allow producers to satisfy demands from society, retailers and consumers, while they are also able to produce high-quality broiler meat in a cost-effective way.

Of course, the focus on broiler welfare varies widely around the globe. While broiler welfare is an important issue in the European Union, countries like the United States, Brazil or China have been less focused on welfare (Robins and Phillips, 2011). Having said that, also food-producing companies in those countries are under pressure from NGOs and society to develop more sustainable methods of meat production. The focus for broiler production around the world should be to breed broilers that can be kept without major health and welfare problems and with low mortality rates, in line with the recommendation of the SCAHAW in 2000 and 2012. Furthermore, broiler management focused on maintaining a good litter quality should be promoted, as this was proven to be pivotal in maintaining good broiler welfare (Dawkins et al., 2004).

7 Future trends in research

Research on innovative hatching environments and on early feeding will likely continue to receive a lot of attention in future years, especially given the fact that benefits are both in the area of broiler health and welfare and human health (through reduced use of antibiotics). This type of research seems key to continue to develop the broiler industry into a more sustainable industry, which uses broilers that are more resilient to environmental changes and that are less susceptible to the health and welfare problems reported in conventional broilers in the past (SCAHAW, 2000; de Jong et al., 2012). Also, more research is needed on alternative and free-range systems. As described previously in this chapter, these middle-market systems seem very promising from both an animal welfare and an economic perspective. It would be interesting to investigate which attributes are most important for broiler welfare: slower-growing genotype, extra space, litter material or outdoor access. When it is clear which attribute or attributes are most important, it could also be investigated how one or more of these attributes could be applied to conventional broiler production systems. The slower-growing genotype seems to play a key role in increasing broiler activity and improving health and welfare. However, this genotype is per definition less efficient in growth than the conventional fast-growing broiler. For future broiler production, it should be investigated how efficiency and welfare can be balanced (Marchant Forde and Rodenburg, 2016). Despite the stabilising and even declining demand for meat in the Western world, demand for animal products is expected to double in the near future, mainly due to increased demand from Asia, the Middle East and Africa. The question is if and how we can meet this expected increased demand for poultry meat in a sustainable way? Should this be by intensifying existing poultry production facilities or by increasing the number of farms worldwide? If mega-farms develop, can we still guarantee good broiler health and welfare on those farms? Scientists should play a central role in this debate and inform the various stakeholders involved on the pros and cons of the various options for broiler health and welfare.

8 Where to look for further information

For further information on broiler welfare, the report 'The welfare of chickens kept for meat production (broilers)' by SCAHAW remains a very good starting point (SCAHAW, 2000). It describes modern broiler production, specific welfare problems in broilers, environmental factors linked to those welfare problems, the role of nutrition and feeding, welfare issues in broiler breeders (not covered in this chapter) and socio-economic aspects. The report was updated in 2012 (de Jong et al., 2012). A second useful resource is *Measuring and Auditing Broiler Welfare* (Weeks and Butterworth, 2003). This edited book first focuses on welfare problems in broilers and then explores how these welfare problems can be measured and audited and how they are linked to environmental factors. Also the 2006 review paper of Professor Werner Bessei on welfare of broilers is a useful source of information, focusing both on genetic and environmental aspects of broiler welfare issues (Bessei, 2006). Regarding the effects of stocking density on broiler welfare, the paper by Professor Marian Dawkins and colleagues is a must-read, also for being one of the few *Nature* papers in the field of animal welfare. For practical welfare assessment of broilers also the Welfare Quality Assessment Protocol for Poultry is very useful, presenting one method of measuring a number of animal- and environment-based welfare indicators and aggregating those into a welfare score that can be compared between flocks and farms (Welfare Quality Assessment Protocol for Poultry, 2009).

9 References

Ask, B. 2010. Genetic variation of contact dermatitis in broilers. *Poultry Science*, 89, 866-75.

Bessei, W. 2006. Welfare of broilers: a review. *World's Poultry Science Journal*, 62, 455-66.

Dal Bosco, A., Mugnai, C., Rosati, A., Paoletti, A., Caporali, S. and Castellini, C. 2014. Effect of range enrichment on performance, behavior and forage intake of free-range chickens. *Journal of Applied Poultry Research*, 23, 137-45.

Dawkins, M. S., Cook, P. A., Whittingham, M. J., Mansell, K. A. and Harper, A. E. 2003. What makes free-range broiler chickens range? In situ measurement of habitat preference. *Animal Behaviour*, 66, 151-60.

Dawkins, M. S., Donnelly, C. A. and Jones, T. A. 2004. Chicken welfare is influenced more by housing conditions than by stocking density. *Nature*, 427, 342-4.

Dawkins, M. S. and Layton, R. 2012. Breeding for better welfare: genetic goals for broiler chickens and their parents. *Animal Welfare*, 22, 147-55.

De Jong, I., Berg, C., Butterworth, A. and Estevéz, I. 2012. Scientific report updating the EFSA opinions on the welfare of broilers and broiler breeders. *EFSA Supporting Publications*, 9, 295E-n/a.

De Jong, I. C., Lourens, A. and Van Harn, J. 2015. Effect of hatch location and diet density on footpad dermatitis and growth performance in broiler chickens. *Journal of Applied Poultry Research*, 24, 105-114.

Fanatico, A. C., Mench, J. A., Archer, G. S., Liang, Y., Gunsaulis, V. B. B., Owens, C. M. and Donoghue, A. M. 2016. Effect of outdoor structural enrichments on the performance, use of range area, and behavior of organic meat chickens. *Poultry Science*, 95, 1980–8.

Gocsik, E., Brooshooft, S. D., de Jong, I. C. and Saatkamp, H. W. 2016. Cost-efficiency of animal welfare in broiler production systems: a pilot study using the Welfare Quality (R) assessment protocol. *Agricultural Systems*, 146, 55–69.

Hocking, P. M. 2014. Unexpected consequences of genetic selection in broilers and turkeys: problems and solutions. *British Poultry Science*, 55, 1–12.

Jacobs, L., Delezie, E., Duchateau, L., Goethals, K., Ampe, B., Lambrecht, E., Gellynck, X. and Tuyttens, F. A. M. 2016. Effect of post-hatch transportation duration and parental age on broiler chicken quality, welfare, and productivity. *Poultry Science*, 95, 1973–9.

Jansson, D. S., Nyman, A., Vagsholm, I., Christensson, D., Goransson, M., Fossum, O. and Hoglund, J. 2010. Ascarid infections in laying hens kept in different housing systems. *Avian Pathology*, 39, 525–32.

Julian, R. J. 1998. Rapid growth problems: Ascites and skeletal deformities in broilers. *Poultry Science*, 77, 1773–80.

Kapell, D. N. R. G., Hill, W. G., Neeteson, A.-M., Mcadam, J., Koerhuis, A. N. M. and Avendaño, S. 2012a. Genetic parameters of foot-pad dermatitis and body weight in purebred broiler lines in 2 contrasting environments. *Poultry Science*, 91, 565–74.

Kapell, D. N. R. G., Hill, W. G., Neeteson, A. M., Mcadam, J., Koerhuis, A. N. M. and Avendano, S. 2012b. Twenty-five years of selection for improved leg health in purebred broiler lines and underlying genetic parameters. *Poultry Science*, 91, 3032–43.

Knowles, T. G., Kestin, S. C., Haslam, S. M., Brown, S. N., Green, L. E., Butterworth, A., Pope, S. J., Pfeiffer, D. and Nicol, C. J. 2008. Leg Disorders in Broiler Chickens: Prevalence, Risk Factors and Prevention. *PLoS ONE*, 3, e1545.

Lamot, D. M., Van de Linde, I. B., Molenaar, R., van der Pol, C. W., Wijtten, P. J. A., Kemp, B. and van den , H. 2014. Effects of moment of hatch and feed access on chicken development. *Poultry Science*, 93, 2604–14.

Marchant Forde, J. N. and Rodenburg, T. B. 2016. Future directions for applied ethology. In Brown, J. A., Seddon, Y. and Appleby, M. C. (Eds), *Animals and Us: 50 years and more of Applied Ethology*. Wageningen, The Netherlands: Wageningen Academic Publishers, pp. 297–317.

Marchewka, J., Watanabe, T. T. N., Ferrante, V. and Estevez, I. 2013. Welfare assessment in broiler farms: Transect walks versus individual scoring. *Poultry Science*, 92, 2588–99.

Pakdel, A., Bijma, P., Ducro, B. J. and Bovenhuis, H. 2005. Selection strategies for body weight and reduced ascites susceptibility in broilers. *Poultry Science*, 84, 528–35.

Powell, D. J., Velleman, S. G., Cowieson, A. J., Singh, M. and Muir, W. I. 2016a. Influence of chick hatch time and access to feed on broiler muscle development. *Poultry Science*, 95, 1433–48.

Powell, D. J., Velleman, S. G., Cowieson, A. J., Singh, M. and Muir, W. I. 2016b. Influence of hatch time and access to feed on intramuscular adipose tissue deposition in broilers. *Poultry Science*, 95, 1449–56.

Rauw, W. M., Kanis, E., Noordhuizen-Stassen, E. N. and Grommers, F. J. 1998. Undesirable side effects of selection for high production efficiency in farm animals: a review. *Livestock Production Science*, 56, 15–33.

Rekaya, R., Sapp, R. L., Wing, T. and Aggrey, S. E. 2013. Genetic evaluation for growth, body composition, feed efficiency, and leg soundness. *Poultry Science*, 92, 923-9.

Robins, A. and Phillips, C. J. C. 2011. International approaches to the welfare of meat chickens. *Worlds Poultry Science Journal*, 67, 351-69.

Rodenburg, T. B., van der Hulst-van Arkel, M. C. and Kwakkel, R. P. 2004. Campylobacter and Salmonella infections on organic broiler farms. *NJAS Wageningen Journal of Life Sciences*, 52, 101-8.

Ruis, M. A. W., Coenen, E., van Harn, J., Lenskens, P. and Rodenburg, T. B. 2004. Effect of an outdoor run and natural light on welfare of fast growing broilers, In: Hanninen, L., Valros, A. (Eds), Proceedings of the 38th Interantional Congress of the ISAE, Helsinki, Finnland, p. 255.

Sandoe, P., Nielsen, B. L., Christensen, L. G. and Sorensen, P. 1999. Staying good while playing god - The ethics of breeding farm animals. *Animal Welfare*, 8, 313-28.

SCAHAW 2000. The welfare of chickens kept for meat production (broilers). European Commission. 149pp. http://ec.europa.eu/food/fs/sc/scah/out39_en.pdf.

Stadig, L. M., Rodenburg, T. B., Reubens, B., Aerts, J., Duquenne, B. and Tuyttens, F. A. M. 2016. Effects of free-range access on production parameters and meat quality, composition and taste in slow-growing broiler chickens. *Poultry Science*, 95, 2971-8.

Thiruvenkadan, A. K., Prabakaran, R. and Panneerselvam, S. 2011. Broiler breeding strategies over the decades: an overview. *World's Poultry Science Journal*, 67, 309-36.

van de Ven, L. J. F., van Wagenberg, A. V., Decuypere, E., Kemp, B. and van den Brand, H. 2013. Perinatal broiler physiology between hatching and chick collection in 2 hatching systems. *Poultry Science*, 92, 1050-61.

van de Ven, L. J. F., van Wagenberg, A. V., Groot Koerkamp, P. W. G., Kemp, B. and van den Brand, H. 2009. Effects of a combined hatching and brooding system on hatchability, chick weight, and mortality in broilers. *Poultry Science*, 88, 2273-9.

van Horne, P. L. M., van Harn, J. J. and Rodenburg, T. B. 2004. Slower growing broilers: performance, mortality and welfare. In Yalcin, S. (Ed.), XXII World's Poulty Congress - Abstract Book, Istanbul, Turkey, 326.

van Overbeke, I., Duchateau, L., de Zutter, L., Albers, G. and Ducatelle, R. 2006. A Comparison Survey of Organic and Conventional Broiler Chickens for Infectious Agents Affecting Health and Food Safety. *Avian Diseases*, 50, 196-200.

Weeks, C. A. and Butterworth, A. 2003. *Measuring and Auditing Broiler Welfare*, Wallingford, United Kingdom, CABI Publishing.

Welfare Quality 2009. Welfare Quality Assessment Protocol for Poultry (broilers, laying hens). Welfare Quality Consortium, Lelystad, The Netherlands. p. 111, http://edepot. wur.nl/233471.

Zuidhof, M. J., Schneider, B. L., Carney, V. L., Korver, D. R. and Robinson, F. E. 2014. Growth, efficiency, and yield of commercial broilers from 1957, 1978, and 2005. *Poultry Science*, 93, 1-13.

Chapter 2

Poultry health monitoring and management: bone and skin health in broilers

Gina Caplen, University of Bristol, UK

1 Introduction

Consumer demand for cheap meat has driven the broiler industry to set ever-increasing production targets, often at the expense of the animals' wellbeing. High levels of lameness and contact dermatitis, an inflammatory skin condition, within intensively reared broiler flocks have long raised welfare concern. Reduced mobility, physical restrictions in behavioural expression and pain experience, associated with leg pathologies and skin lesions, will have negative welfare repercussions. Despite some success with interventions to improve bone and skin conditions, including sustained efforts by the genetic companies to select for improvements in leg health, some of the same problems still persist decades later. One estimate indicated that 12.5 billion broilers experience leg problems annually worldwide (FAO, 2010).

With demand for chicken meat still escalating globally, it is imperative that the risk factors for lameness and contact dermatitis are fully understood so that novel and effective management solutions can be developed and better employed. Not only would the improvement of leg and skin health directly benefit broiler welfare, but there is also potential to confer many economical benefits to the producers. Lameness is associated with higher flock morbidity and mortality (Dinev, 2009) and is one of the primary causes of on-farm culling (especially in older birds). In addition, both lameness and contact dermatitis lower growth performance indices (Knowles et al., 2008; Butterworth

http://dx.doi.org/10.19103/AS.2020.0078.17

and Haslam, 2009) and are responsible for a large proportion of carcass condemnations at the slaughter plant (Kyvsgaard et al., 2013; Hashimoto et al., 2013).

2   Leg disorders and lameness

2.1 Categories of lameness and impaired walking ability

Leg disorders encompass a broad range of pathologies and can involve any combination of tendons, joints, ligaments and bones (Bradshaw et al., 2002). Lameness can arise due to developmental bone deformities, be of an infectious origin, or be degenerative (e.g. develop as a consequence of a trauma or progressive load-bearing).

2.1.1 Gait and morphology

Over the last 60 years the morphology, walking style and locomotive capacity of commercial broilers have been profoundly altered by genetic selection for increased growth rate and body mass. Selection has specifically targeted the breast muscle, which, as a proportion of the total body mass, is now double that of traditional strains (Schmidt et al., 2009). This 'pectoral hypertrophy', in addition to greater thigh muscle and leg bone mass, and relatively short legs distinguish modern broiler strains from traditional meat breeds.

Broilers display pronounced gait alterations when compared to red junglefowl (the ancestral species), and additional differences are evident between non-lame and moderately lame individuals (Caplen et al., 2012). Compensatory gait adaptations, including the wider stance and more pronounced hip and foot rotations characteristic of broilers, develop in response to increased load (body mass), morphology (unbalanced posture), pathology and injury (Corr et al., 2003a,b). In the short term such adaptations minimise the additional energy required for movement and enable the birds to retain mobility; however, ultimately, the gait adaptations themselves are likely to be fatiguing and increase the risk of modern broilers developing progressive leg pathologies.

2.1.2 Developmental lameness

A rapid growth rate and unnaturally high body mass place extreme loads on immature bones and joints. Frequently observed skeletal abnormalities include rotational (tortional) deformities such as 'twisted legs' (unilateral angulation of tibiotarsal articulation) and 'valgus–varus' deformities (Riddell and Kong, 1992). Valgus or varus angulation is mainly associated with bilateral tibial deviation and rotation (often with gastrocnemius tendon displacement); however, some

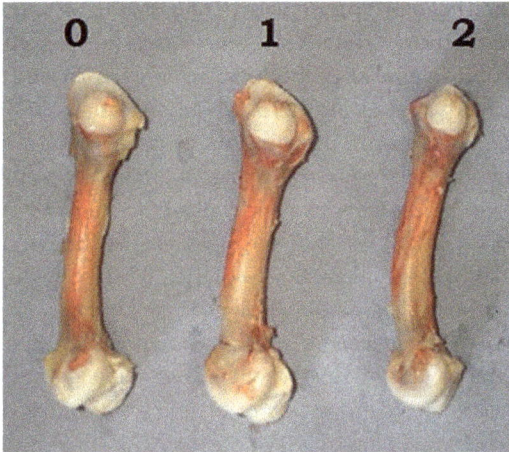

Figure 1 Examples of femoral curvature: 0 = normal, almost straight; 1 = mild curvature; 2 = moderate curvature (photo courtesy of Dr Andrew Butterworth, University of Bristol).

studies also report femoral deformations (e.g. Duff and Thorp, 1985). Such deviations remain poorly understood but are likely to be a consequence of poor bone mineralisation. Leg angulations and bone deformities can occur over a range of severities; for example valgus has been described as mild (tibia-metatarsus angle between 10° and 25°), intermediate (25-45°), and severe (>45°) (Shim et al., 2012a). A severity scale for femoral curvature is illustrated in Fig. 1.

Tibial dyschondroplasia (TD) is a leg pathology resulting from inadequate ossification and vascularisation of the epiphyseal growth plates. The TD lesion is characterised by an abnormal mass of cartilage occurring, most commonly, in the tibial metaphysis (Orth and Cook, 1994). Severe TD lesions appear to cause lameness, although the effect is less evident in milder forms (Lynch et al., 1992). Rickets is characterised by an enlargement of the epiphysial growth plate and soft rubbery bones, a result of inadequate endochondral ossification and mineralisation (Julian, 1998; Dinev and Kanakov, 2011). Rickets develops as a result of nutrient malabsorption, for example following intestinal disease, or due to inadequate nutrition, and is relatively uncommon within modern broiler flocks.

2.1.3 Infectious lameness

Bacterial chondronecrosis with osteomyelitis (BCO), more specifically known as femoral head necrosis, is recognised as a major cause of lameness in commercial broilers worldwide and is reviewed in detail elsewhere (e.g. Wideman, 2016). BCO is primarily identified by the presence of lesions in the femur (in particular, the metaphysis, femoral head and proximal femoral growth plate) and the tibiotarsus. The condition is associated with a complex range of opportunistic infective organisms, predominantly avian pathogenic *Escherichia coli* and *Staphylococcus* spp. (Dinev, 2009; Mandal et al., 2016; Al-Rubaye et al., 2017; Wijesurendra et al., 2017). Rapid increases in body weight cause mechanical damage to the immature cartilage (osteochondrosis). Bacterial proliferation at the wound sites triggers an immunological response and limits blood flow to the growth plate (Wideman and Prisby, 2013), which, in turn, leads to the development of necrotic abscesses and voids that are characteristic of BCO (Wideman, 2016). Bacteria are thought to be distributed haematologically, via the respiratory or gastrointestinal tract; *S. agnetis*, for example, can be transferred to the blood stream via drinking water (Al-Rubaye et al., 2017). Since broilers favour sitting, reduced blood flow resulting from prolonged compression of the leg arteries may further promote the development of this pathology (Wideman, 2016). Broilers housed on wire flooring are particularly susceptible to developing BCO lesions due to persistent footing instability and physiological stress (Al-Rubaye et al., 2017).

Tenosynovitis (an arthritis caused by avian reoviruses) causes the inflammation of the hock joints, lesions in the gastrocnemius and digital flexor tendons and, ultimately, lameness (Sellers, 2017). The gastrocnemius tendons leave the gastrocnemius muscle and pass over the intertarsal (hock) joint and attach to the posterior surface of the tarsometatarsus, while the digital flexor tendons extend along the tarsometatarsus to the phalanges. Upon dissection the infected synovia of swollen hocks often appears thickened, often in association with an increase in joint fluid, blood or pus (Fig. 2). In China, bacterial arthritis is also common in broilers, the main serovar being *Salmonella pullorum* (Guo et al., 2019).

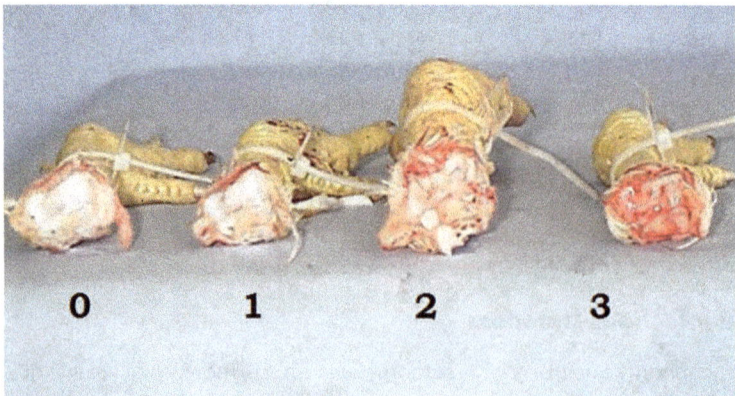

Figure 2 Inflammation of the joint capsule (proximal tarsometatarsus). Synovitis score scale: 0 = none, 1 = mild, 2 = medium, 3 = severe (photo courtesy of Dr Andrew Butterworth, University of Bristol).

2.2 Qualitative and quantitative assessment of leg health

Several gait scoring methods, based upon the visual appraisal of walking ability, have been developed for assessing broiler lameness. The widely used Bristol six-point 'gait score' (GS) scale, developed by Kestin et al. (1992), has now been adopted in a modified form for use within the broiler Welfare Quality® Assessment; scores range between 'GS 0: Normal, dextrous and agile' and 'GS 5: Incapable of walking' (Welfare Quality, 2009). Although this system is suited for on-farm welfare assessment (being relatively quick and requiring no specialised equipment), it lacks the capacity to discriminate between lameness types. A moderately lame bird assigned 'GS 3' could be affected bilaterally (e.g. valgus) or unilaterally (e.g. singular hock inflammation), or lack any obvious pathology.

Objective methodologies for quantitatively assessing gait are available, including kinetic (the measure of forces involved in walking) and kinematic (the study of body motion) systems, but at the current time these technologies remain better suited to experimental work. Several techniques have been employed to collect kinetic data from broilers, with varying measures of success. These include force plates (Corr et al., 2007; Sandilands et al., 2011), which require a constant walking speed and generate significant background 'noise', and piezoelectric pressure-sensing mats (Nääs et al., 2010). Image analysis systems have enabled the collection of three-dimensional kinematic data for the purpose of correlating gait characteristics, such as walking speed, step length and lateral body oscillation (Fig. 3), with defined lameness scores, (Caplen et al., 2012; Aydin, 2017a) and have been used to detect subtle localised changes in gait parameters as part of analgesic drug studies (Caplen et al., 2013a).

The latency-to-lie (LTL) test has been widely employed as a simple index of leg weakness (Weeks et al., 2002; Berg and Sanotra, 2003). The test is based

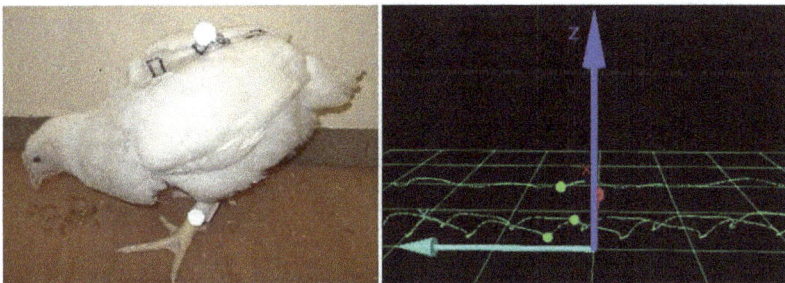

Figure 3 Broiler fitted with retro-reflective markers (one on the midline of the back, and a further two attached to the posterior aspect of the metatarsal bone, immediately above each foot). An infra-red four-camera motion capturing system was used to collect three-dimensional kinematic data as the bird walked down a runway. Images: Gina Caplen, University of Bristol.

upon the premise that chickens find sitting in water aversive and that birds with poor leg health will sit sooner when they are placed standing in shallow water (Fig. 4). Due to the logistics involved in performing this test, it is better suited to experimental trials rather than on-farm welfare assessments. An automated visual monitoring methodology has been recently developed to assign GS to broilers based upon 'the number of lying events' and 'latency to sit' when birds traverse a test corridor (Aydin et al., 2015; Aydin, 2017b). Although this new system may have the potential for transfer to a farm environment, any methodology that relies upon broilers voluntarily moving along a standard walkway will run the risk of sample bias towards the more mobile individuals (and under sample the lame birds) within a flock.

Significant efforts are currently being made to develop automated farm-based camera systems that remotely monitor flock behaviour and forewarn producers of the development of muscular-skeletal problems, for example when flock activity levels fall below an accepted level. Image analysis systems are reviewed in Chapter 7. In brief, several systems show great promise for incorporation with existing management software and wide-scale employment. Optical flow measures of flock movement have been shown to correlate with lameness (Dawkins et al., 2009, 2013; Roberts et al., 2012), while a different visual system has also been used to correlate bird activity and flock distribution with GS (Van Hertem et al., 2018).

Infrared thermography (IRT) provides a non-invasive means of measuring infrared radiation (surface heat) from an object, and the technology is being

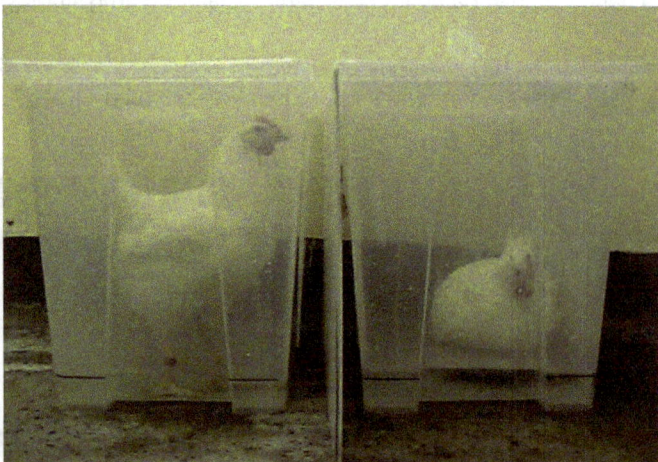

Figure 4 The latency-to-lie test. Broilers are placed standing into shallow water at room temperature (without visual contact of conspecifics) as per the left-hand bird. The time taken for the birds to sit, as per the right-hand bird, provides a simple index of leg weakness. Birds are removed from the water as soon as they sit. Image: Gina Caplen, University of Bristol.

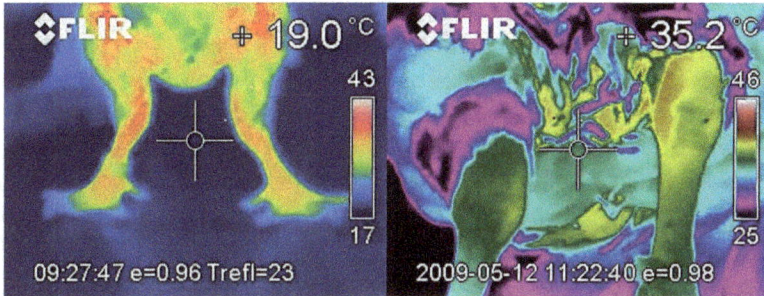

Figure 5 Thermal images of a broiler with a developmental 'valgus' leg deformity (left image) and a broiler with unilateral inflammation of the right hock joint (right image). Images: Gina Caplen, University of Bristol.

increasingly utilised in the fields of clinical and veterinary health. Although IRT cannot diagnose specific pathologies, it can be used to detect localised areas of increased or decreased heat production, that is due to inflammation or reduced blood flow, accordingly (Fig. 5). Relating to poultry production, IRT may prove to be useful as a non-invasive technique for detecting and monitoring leg pathologies on farm that are currently diagnosable only at post-mortem, for example BCO (Weimer et al., 2019).

2.3 Prevalence of lameness and specific leg pathologies

According to a comprehensive UK study, by the end of the rearing period more than a quarter (27.6%) of standard broilers had a moderate to severe gait impediment (GS 3+) and 3.3% were unable to walk (Knowles et al., 2008). More recent studies estimate a slightly lower prevalence of moderate to severe gait impediments in European flocks (Europe-wide: 15.6%, Bassler et al., 2013; Norway: 24.6%, Kittelsen et al., 2017; 19%, Granquist et al., 2019). A Danish survey found 77% of conventional broilers to have at least some form of impaired walking ability (GS 1+), while the prevalence of birds with moderate to severe lameness (GS 2+) was, a modest, 6% (Tahamtani et al., 2018).

Very little information is available regarding the current prevalence of specific leg-health pathologies. Sanotra et al. (2003) report very high levels of TD within Scandinavian flocks (Denmark: 57.1%; Sweden: 45.2-56.3%), yet later studies report much less (Finland: 2.3%, Kaukonen et al., 2017; Denmark: 5%, Tahamtani et al., 2018). BCO appears to retain a persistent presence within broiler flocks. BCO lesions have been recorded in >90% of lameness-related mortality cases (Bulgaria: Dinev, 2009), and in almost 30% of all broiler mortalities and culls (Australia: Wijesurendra et al., 2017). Although joint lesions (including arthritis and tenosynovitis) were reported to increase within UK flocks by 35% between 2011 and 2013, the actual increase was relatively small (2.16

to 2.92 per 10 000 birds slaughtered, Part et al., 2016) and, therefore, perhaps not fully reliable. Valgus and other long-bone deformities are frequently seen within standard UK broiler flocks, but exact estimates are not available. An earlier study reports a very high prevalence of valgus–varus in Scandanavia (Denmark: 36.9%; Sweden: 46.4–52.6%, Sanotra et al., 2003).

A wide variation in the estimates of general walking ability and specific leg disorders is to be expected when you consider the diversity in production systems employed on both regional and global scales (including stocking density, climate, barn design and diet). The different housing systems for broilers are described in Chapter 10. In addition, breeding companies are constantly selecting for improved production parameters and leg health (e.g. valgus/varus and TD), and so the genotype of birds supplied for use in commercial systems is in perpetual flux. In addition, management practices can become out-dated. For example, vaccination had successfully controlled tenosynovitis for decades; however, in recent years there has been a dramatic increase in the number of clinical cases as new genetic virus variants emerge (Sellers, 2017). There is an obvious need for independent and systematic monitoring of specific pathologies to document and better understand such trends.

2.4 Lameness risk factors

Key risk factors associated with lameness and poor leg health in modern broilers include those directly associated with the growth rate (including genotype, age, body mass and feeding regimen) and those indicative of sub-optimal environmental management.

2.4.1 Sex, age and body mass

Lameness increases with broiler age and body mass (e.g. Henriksen et al., 2016). Males generally have a higher GS than females and are more prone to developing valgus–varus and femoral degenerative joint lesions (Paz et al., 2013), especially if they are heavy. Some studies report a higher TD incidence in male broilers (e.g. Birgul et al., 2012), presumably since TD is also positively associated with body mass (e.g. Shim et al., 2012a). Worryingly, TD lesions have been recorded in broilers as young as 20 days, which may be a direct consequence of selection pressure for increased developmental rates (Dinev et al., 2012).

2.4.2 Genotype and system

Slower-growing genotypes demonstrate less lameness than modern fast-growing breeds when reared on the same feeding regimen (Kestin et al., 2001). Fast-growing breeds tend to have lower bone strength, which makes them susceptible to developing leg deformities (Shim et al., 2012b). The choice of system (and appropriate broiler genotype) has an obvious impact upon the leg health of the flock. Organically reared broilers tend to have better leg health than conventional flocks (Tuyttens et al., 2008; Tahamtani et al., 2018); however, flock size and stocking density are usually much lower in the organic farms, and organic systems are required to use slower-growing breeds. Fast-growing breeds do not perform well under extended production and are unsuited for use in extensive systems, as the birds make low use of the outdoor space and have a tendency to develop severe welfare problems, including high cull and mortality rates, impaired mobility, joint inflammation and severe footpad dermatitis (FPD) (Nielsen et al., 2003; Dal Bosco et al., 2014; Castellini et al., 2016).

2.4.3 Stocking density

Research consistently indicates that the health and welfare of broilers is compromised above stocking densities of 34–38 kg/m², dependent upon final body weight (e.g. Estevez, 2007; Sun et al., 2013; Das and Lacin, 2014). Accordingly the European Broiler Directive (2007/43/EC), which lays down the minimum rules for the protection of chickens kept for meat production, specifies that, as a general rule, stocking density should not exceed 33 kg/m². Although broilers grow more slowly at high stocking densities (Dawkins et al., 2004; Sun et al., 2013), there is evidence that leg health can become compromised at densities as low as 23 kg/m² (Buijs et al., 2009). Birds present on the barn floor effectively act as obstacles, and this has been termed the 'barrier effect' (e.g. Collins, 2008). Prolonged activity becomes more difficult at higher stocking densities, and behaviour becomes fragmented; locomotion slows and walking bouts decrease in length (e.g. Buijs et al., 2011; Ventura et al., 2012). Although stocking density does not appear to be directly linked with the prevalence of specific pathologies (e.g. valgus–varus: Arnould and Faure, 2004; TD: Das and Lacin, 2014), higher stocking densities are associated with tibial deformities and reduced bone strength (Buijs et al., 2012; Sun et al., 2013; Vargas-Galicia et al., 2017). The apparent effects of stocking density on skin health are due to correlations with deteriorating environment, especially litter quality, and this is associated with poor environmental control (Section 3.3.1). Optimal environmental management is essential for maintaining general flock health status since the combination of wet litter with warm moist air will promote bacterial growth and transmission. Gut infections, such as enteritis, provide one of many indirect mechanisms by which the intestinal microbiota may influence skeletal fitness and bone mass (Charles et al., 2015), while lameness may also arise directly as a result of bone or joint infection.

2.4.4 Environmental conditions

Key requirements for achieving and sustaining good leg health include adequate ventilation of the production facility, provision of high-quality litter and the maintenance of temperature and relative humidity (RH) within an optimum range. Higher GSs have been associated with increased hock burn (HB) and FPD scores (see Section 3), and reduced feather cleanliness, suggesting that a sub-optimal physical environment (i.e. poor litter quality) may be detrimental to leg health (Granquist et al., 2019). High temperature, high ammonia concentrations and the percentage of time that temperature and RH remain outside of the breeder's recommended range have also been associated with gait deficiencies and leg deformities (Dawkins et al., 2004; Jones et al., 2005; Tullo et al., 2017).

Rigid environmental control is also very important during incubation; temperature and oxygen concentrations can influence embryonic bone development, post-hatch production parameters and leg health (e.g. Ipek and Sozcu, 2016; Oznurlu et al., 2016). TD incidence has been associated with temperature deviations during the early stages of embryo development (Yalçin et al., 2007). Small changes in egg shell temperature (EST) during incubation could easily occur in practice due to the widespread use of large-capacity incubators and the fast growth rate of broiler embryos. Indeed, Oviedo-Rondón et al. (2009a) observed that broilers hatched from multi-stage incubators went on to develop a higher GS and greater prevalence of valgus leg deformations than those hatched from single-stage incubators.

2.5 Welfare impact of lameness

The terms 'lameness' and 'leg health' are not fully interchangeable. Although correlations between body mass, growth rate and lameness are well documented (e.g. Kestin et al., 2001), there is little evidence to conclusively link lameness severity with pathology (Garner et al., 2002; Sandilands et al., 2011; Fernandes et al., 2012). Instead, lameness is more likely to reflect the birds' subjective experience of the pathology, the manifestation of an integrated behavioural response. Investigating the true impact of poor leg health at the bird level is complicated by the gross morphological modifications that define the modern broiler bird and the multiple, often inter-linked, aetiologies and pathologies to which they are susceptible. As already mentioned lameness has been associated with contact dermatitis, and it is likely that, if left untreated, many progressive leg pathologies will confer a greater risk of developing secondary complications. Condemnations (carcasses deemed unfit for human consumption) at post-mortem inspection have been associated with increasing GS (Granquist et al., 2019), indicating that at least a proportion of lame broilers display pathological changes on the carcasses.

2.5.1 Dehydration and reduced feeding

Lame broilers may endure dehydration (Butterworth et al., 2002) since decreased mobility can limit their ability to access water, even in systems that otherwise provide appropriate water delivery. Nipple drinkers minimise water spillage and, as such, can help to maintain litter quality; however, their position, above the broilers' heads, can cause some birds to lose balance when stretching to drink (Jones et al., 2005). Very lame birds are often seen to have a lower body mass. Food intake may be reduced if it is difficult to reach; however, appetite may also become suppressed as a result of inflammatory-induced sickness behaviour (e.g. Dantzer, 2009).

2.5.2 Pain

The prevalence and severity of pain associated with broiler lameness remains poorly understood due to the impact of biomechanical factors on gait pattern, the heterogeneous nature of leg pathologies, and a requirement for rigorous experimental design (since pain experience must be inferred using indirect non-verbal measures). Strong physiological evidence indicates that broilers have the capacity to experience leg pain. Slowly adapting mechanoreceptors are present within the skin of the chicken tarsometatarsus, and these become sensitised following induced inflammation (Gentle et al., 2001). Inflammatory arthropathy, a condition that can cause pain in humans, has also been identified within the hock joints of spontaneously (naturally) lame broilers (Corr et al., 2003c). Some very lame broilers have also been reported as having a greater relative adrenal mass, which is likely to be indicative of chronic stress (Müller et al., 2015).

Visual identification of a pain state is complicated by the fact that non-lame broilers spend the majority of their time sitting (Weeks et al., 2000). In addition, poultry are a prey species and it is likely that they avoid overt pain-associated behaviour, as any display of weakness could make them more vulnerable to predation. Morphology (via mechanical limitations and a reduced motivation to walk) has a significant influence upon gait and activity levels in modern strains, regardless of any assumed discomfort. Since certain pathologies (e.g. a mild skeletal deformity) are likely to be less painful than others (e.g. inflammatory or necrotic conditions), the welfare implications of failing to quantify and differentiate pain from other causes of gait abnormality are substantial.

The provision of analgesic drugs under experimental conditions can provide indirect evidence for pathological pain if positive changes in behaviour, or improvements in a predefined test performance, are observed post-treatment. The majority of studies to date have utilised non-steroidal anti-inflammatory drugs (NSAIDs), such as carprofen and meloxicam, as they are routinely used to manage pain associated with osteoarthritis in dogs and cats and may have a therapeutic potential in poultry. A self-selection experiment, reporting the preferential selection of food spiked with NSAID, claimed to provide early evidence for the occurrence of pain in lame broilers (Danbury et al., 2000); however, a later study was unable to corroborate these findings (Siegel et al., 2011).

NSAID treatment has been observed to increase walking velocity and modify gait in moderately lame broilers (Nääs et al., 2009; Caplen et al., 2013a) and improve LTL performance (Hothersall et al., 2016). However, such improvements in mobility could be attributable to a reduction in joint inflammation as opposed to any direct analgesic effect. Although the relationship between lameness and pain (and thus welfare) is complicated by confounding 'risk factors' such as sex, strain, bodyweight and pathology, 'lameness' has been identified as the most consistent predictor for several broiler mobility measures. This indicates that there is a constituent of 'lameness' that cannot be explained by any combination of the more obvious bird characteristics (e.g. being a heavy male), and it is this component that may represent pain or discomfort (Caplen et al., 2014).

Hyperalgesia (heightened sensitivity to pain) is prevalent in many disease states as part of an inflammatory response to prevent further tissue damage. Pain-producing chemicals (cytokines and chemokines) trigger primary afferent nociceptor sensitisation. Primary hyperalgesia describes pain sensitivity that occurs directly in the damaged tissues; a lowered nociceptive/pain threshold to both thermal and mechanical stimuli is usual. Caplen et al. (2013b) utilised a specially designed apparatus to detect primary thermal hyperalgesia in broilers with experimentally induced inflammatory arthropathies (an acute pain model) and demonstrated that NSAID treatment could reverse this effect via anti-nociception (Fig. 6). A comparable effect has not yet been demonstrated in broilers with non-induced 'spontaneous' lameness; conversely, a higher baseline thermal threshold was reported in farm-lame broilers, which increased further following NSAID treatment, potentially due to altered nociceptive processing (Hothersall et al., 2014). The fact that hyperalgesia was detected in a group of experimental birds following pain induction under controlled conditions but not in a group of chickens with (potentially mixed) on-farm pathologies does not prove that the latter group had no pain experience. Pain is not always accompanied by hyperalgesia, and, therefore, a lack of response is not an evidence for an absence of pain. Clear difficulties exist in obtaining large groups of birds with comparable pathologies for experimental work and in linking lameness severity with pathology (e.g. Sandilands et al., 2011).

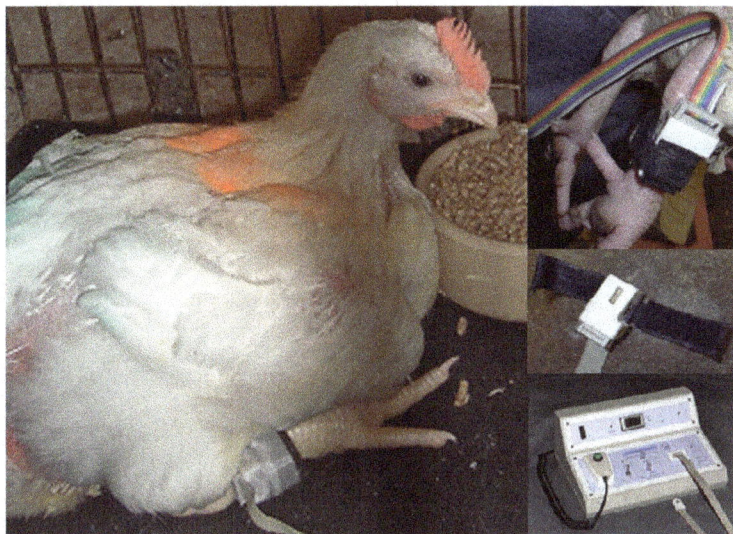

Figure 6 Apparatus used for thermal threshold testing (Images: Gina Caplen). The leg probe (containing a temperature sensor and a heated element) was attached to the lateral aspect of the tarsometatarsus via a Velcro strap. A ramped heat stimulus was applied until a behavioural endpoint was detected and then immediately removed. The response temperature was held on a digital readout.

2.5.3 Limited behavioural expression

Selection for broiler growth has significantly narrowed their ethogram and altered their time budget, compared to other breeds of *Gallus gallus domesticus,* by restricting the range of behaviours that they are physically able to perform. The effect of posture on resting metabolic rate becomes increasingly significant as broilers grow; locomotion becomes very energetically expensive, standing more so than sitting. Since the metabolic scope for exercise decreases throughout their development, the proportion of the overall metabolic rate accounted for by locomotor behaviour also decreases, which corresponds to declining activity levels and low walking speeds (Tickle et al., 2018). This is particularly apparent in the fast-growing breeds, since slower-growing broilers perform less sitting and more perching, walking and ground scratching throughout the production period (Bokkers and Koene, 2003; Reiter and Bessei, 2009).

In welfare terms inherent inactivity is problematic for several reasons. Walking acts to strengthen muscles and bones, and inactivity is thought to be a direct cause of leg weakness. Lameness further compromises mobility, limiting behavioural expression and reducing the value of any enrichment provision. There is evidence that physically impaired individuals retain at least some motivation to perform locomotory behaviour (Rutten et al., 2002; Bokkers and Koene, 2004; Bokkers et al., 2007). When this motivation remains unfulfilled, it is likely to trigger frustration (e.g. as indicated by displacement preening, Bokkers and Koene, 2003), stress and suffering.

2.6 Prevention and control of lameness

2.6.1 Genetics

Long-term selection for improved broiler leg health (in particular, long-bone deformities and TD) over the last 30 years or so has been partially effective, despite some unfavourable genetic correlations with body mass (Kapell et al., 2012a). A marked reduction in TD within commercial strains over recent years is testimony to what is possible. Estimated heritabilities of non-infectious skeletal disorders (including TD and valgus–varus), and susceptibility to infection (e.g. FHN), indicate that genetic selection and breeding programmes offer a means to further reduce multiple lameness aetiologies and leg pathologies alongside the main selection focus, to further improve production parameters (Akbaş et al., 2009; Wideman et al., 2014). At the producer level, a greater commercial uptake of slower-growing strains (with their lower predisposition for musculo-skeletal health problems) housed within appropriate systems would directly alleviate the main welfare concerns.

2.6.2 Incubation conditions

Incubation lighting and heating schedules both appear to have a marked effect upon leg health. Egg exposure to continuous light has been shown to have a detrimental effect upon embryonic leg bone development. The risk of both poor bone strength and developing TD in later life is increased, compared to incubation regimes that include a continuous period (at least 8 h) of darkness (van der Pol et al., 2017, 2019).

The optimum incubation EST for healthy leg development appears to be lower than 37.8°C, the temperature currently recommended for maximising hatchability and chick growth. Embryos incubated under 'slow start' conditions, that is under lower temperatures (EST: 36.9-37.5°C) during the first few weeks of incubation, hatched later, had greater femoral bone ash (Groves and Muir, 2014; Muir and Groves, 2018), grew slower over the first week post-hatch, and had a lower prevalence of TD lesions at day 34 (Groves and Muir, 2017). Hatchery and chick quality issues clearly influence the susceptibility of broilers to BCO. Although it remains widespread within European flocks, the pathology is being successfully addressed via improvements in hatchery hygiene (Dinev, 2009).

2.6.3 Barn management

As mentioned previously, it is very important to effectively utilise heating and ventilation systems to stringently maintain the environmental parameters of the barn within the guidelines for the breed. Maintaining dry litter also appears to be important; the use of wheat straw as litter has been associated with a higher prevalence of lameness than (more absorbent) wood-shavings (Su et al., 2000), while the addition of vermiculite to wood-shavings has been found to lower lameness severity further (Yildiz et al., 2014).

2.6.4 Diet and feeding regimes

Diet and feeding regimes are utilised to control growth rate and/or increase bone strength. Quantitative or qualitative dietary restriction can be used as a means to slow weight gain and promote healthier anatomical leg development (Kestin et al., 2001; Wijtten et al., 2010). However, feed restriction in any form will trigger overt hunger, especially in the meat breeds that have been selected for high appetite and metabolic growth. Hunger is often associated with frustration, stereotypies and adverse behaviour, occasionally including cannibalism (Eriksson et al., 2010), all of which have obvious adverse impacts upon bird welfare.

A change from ad-libitum feeding to meal feeding (i.e. limiting food availability to 240 min/day split between two, three and four discrete meal times) has been seen to improve walking ability and reduce TD (Su et al., 1999). Responsive nutritional management can also be used to control lameness by slowing down the rate at which young broilers grow and then speeding the growth rate back up to achieve a standard finishing commercial body mass once the leg bones have better developed. This can be achieved via the sequential feeding of two diets, a high-energy/low-protein diet and a low-energy/high-protein diet, over 48-hour feeding cycles (Leterrier et al., 2008); a standard finishing diet is provided from day 29 until the last day of production to compensate for reduced growth. The inclusion of whole wheat within the diet is beneficial as it slows digestion and lowers the feed conversion rate, thereby, reducing both growth rate and lameness (Knowles et al., 2008). All three of these management practices, via health improvements and decreased feed costs, have been identified as having economic potential to realise substantial improvements in gross margin and net return for the farmer (Gocsik et al., 2017).

Broilers fed mashed diet have been shown to have higher bone ash and lower GS than broilers fed the same diet in pellet form (Brickett et al., 2007). Presumably birds eat a greater volume of pelleted food (and achieve a higher corresponding weight gain), while those fed mashed diet benefit physiologically from being able to preferentially select components of their ration and benefit behaviourally from spending more time foraging for their feed. Since deficiencies and imbalances of numerous vitamins and minerals can impact upon broiler bone mineralisation and leg health, the provision of an optimal dietary formulation is extremely important. Due to the ongoing selection for improved production parameters, dietary guidelines require a regular review. Numerous nutritional factors influence broiler leg health, often via complex interactions (e.g. facilitating mineral assimilation), and these are reviewed in detail elsewhere (see Waldenstedt, 2006; Kwiatkowska et al., 2017). Calcium (Ca), phosphorus (P) and vitamin D_3 (cholecalciferol) appear to be of key importance.

Dietary Ca supplementation is beneficial for increasing bone quality and reducing TD incidence (Coto et al., 2008; Abdulla et al., 2017); sources such as oyster shell, snail shell and limestone are reported as being most effective (Oso et al., 2011). In addition to quantity, the dietary balance between available Ca and P is extremely important. The optimum feed ratio is 2:1, and an excessive supply of either element can lessen the assimilation of both (Coto et al., 2008; Bradbury et al., 2014). Rickets can occur as a Ca-deficiency or P-deficiency type, resulting from either a direct dietary deficiency or an excessive proportion of either. An increase in TD incidence has also been reported as a consequence of Ca:P imbalance (Waldenstedt, 2006). Vitamin D_3 is beneficial as a dietary supplement since it increases Ca and P intestinal absorption, improves bone quality and walking ability (Baracho et al., 2012; Sun et al., 2013) and is valuable in the prevention of TD (Whitehead et al., 2004) and BCO (Wideman et al., 2015). Ascorbic acid (vitamin C) supplementation has been found to benefit both bone quality (Yildiz et al., 2009) and walking ability (Petek et al., 2005).

Phytase is an enzyme that acts to increase the bioavailability of various minerals (e.g. Ca, P, Mg, Zn, Fe), a large proportion of which are present within grains and seeds as insoluble complexes. Dietary phytase has been shown to increase the apparent digestibility of Ca by 4-6% (Saima et al., 2009) and to improve bone mineralisation (Bradbury et al., 2017).

Dietary supplementation of probiotics (live organisms intended to improve the gut microflora), prebiotics (non-digestible feed components that promote the growth of beneficial intestinal microorganisms) and synbiotics (the combination of a probiotic with a prebiotic) benefit bone development and mineralisation by increasing the intestinal absorption and assimilation of nutrients and minerals (Scholz-Ahrens et al., 2007; Yan et al., 2018). The inclusion of a dietary synbiotic (containing a prebiotic and a probiotic mixture of four microbial strains) was found to improve multiple indices of broiler leg health, LTL performance and walking ability (Yan et al., 2019). Probiotics are most effective at reducing BCO incidence if they are given proactively as part of the feed, rather than therapeutically after the onset of lameness (Wideman et al., 2012).

It is also important to recognise that there are environmental contaminants that can negatively impact upon health, even with the provision of an 'optimum diet'. Bacterial, viral and parasitic infections can reduce the ability of the intestinal epithelium to absorb nutrients, while feed contamination with certain mycotoxins can induce or exacerbate skeletal problems due to interference with vitamin D metabolism (Waldenstedt, 2006).

2.6.5 Photoperiod and light intensity

Continuous bright light is normally provided during the first 4 days following barn placement post-hatch (the brooding period) to stimulate feeding; however, exposure to intermittent lighting during the same period has been shown to benefit leg health by slowing bone development and increasing leg bone symmetry (van der Pol et al., 2015). Within the EU the Broiler Directive (2007/43/EC) states that lighting must follow a 24-h rhythm and include periods of darkness lasting at least 6 h in total, with at least one uninterrupted period of darkness of at least 4 h, excluding dimming periods (see Chapter 10). The presence of a dark phase (scotophase) throughout the production cycle is conclusively beneficial for broiler leg health. Provision of a scotophase is associated with a reduction in lameness (Brickett et al., 2007; Knowles et al., 2008; Bassler et al., 2013; Schwean-Lardner et al., 2013; Das and Lacin, 2014) and TD (Petek et al., 2005; Karaarslan and Nazligul, 2018) and an increase in tibial strength (Lewis et al., 2009; Yang et al., 2015), compared to exposure to continuous light. Bone mineralisation peaks during the dark period and is sensitive to diurnal rhythm (Russell et al., 1984, as cited by Bassler et al., 2013). In addition, broilers provided with a scotophase are physically more active during the light period than those kept under near-continuous light (Sanotra et al., 2002; Bayram and Özkan, 2010; Schwean-Lardner et al., 2012). A photoperiod of 16L:8D appears optimal for maximising both welfare and feed conversion (Classen, 2004).

Broilers are typically housed under low light intensity during the light phase for the majority of the production cycle. Those reared under very dim lighting (<5 lux) develop a higher body mass, display less activity and have a higher GS than birds reared under brighter (10–320 lux) lighting regimes (Blatchford et al., 2009, 2012; Senaratna et al., 2016; Fidan et al., 2017). Exposure of chicks to low levels of UV-B illumination (to simulate natural daylight) has been trialled as an alternative non-dietary means of enhancing vitamin D_3. UV-B treatment has been found to have substantial benefits for bone characteristics and virtually eliminates growth-plate abnormalities caused by Ca and vitamin D deficiencies (Fleming, 2008). Broilers reared under environments including supplementary UV-A light-emitting diode (LED) and UV-B fluorescent lighting had a lower GS than those of a white LED control group (James et al., 2018). The incorporation of UV wavelengths within commercial lighting regimes may, therefore, be beneficial for broiler leg health.

2.6.6 Environmental enrichment

Increasing activity levels have known benefits for broiler leg health since mechanical loading is essential for maintaining normal bone formation and remodelling (Hester et al., 2013). Broilers encouraged to walk on treadmills were shown to have better bone density, thickness and fewer leg bone deformities than control birds (Reiter and Bessei, 2009), while cage-reared broilers were observed to have lower leg bone mineralisation than birds reared in open barn systems (Aguado et al., 2015).

Standard broiler breeds are typically barn-reared in low complexity environments with minimal enrichment; however, studies that have provided commercial flocks with elevated structures in an attempt to encourage locomotor activity (and thus improve leg and foot pad health) have met with varying success. Although the provision of platform perches in combination with dust baths was found unsuccessful in reducing lameness (Bailie et al., 2018a), the provision of platforms in a different study improved GS and TD incidence/severity (Kaukonen et al., 2017). Presumably the key to use is in the platform design. A small-scale study using experimental pens reported that slow-growing breeds use perches more frequently than fast-growing strains (Bokkers and Koene, 2003), presumably due to intrinsically higher activity levels and superior leg strength; however, low usage overall suggests perch provision to be inappropriate for broilers. In contrast, the provision of straw and/or wood-shaving bales has reliably been associated with increased physical activity (Bailie and O'Connell, 2015; Ohara et al., 2015; de Jong and Gunnink, 2019) and improved leg health (Bailie et al., 2013; Vasdal et al., 2019) in commercial flocks.

Body mass appears to physically limit the utilisation of elevated structures. Usage undergoes a marked decline towards the end of the production cycle (perches: Bailie and O'Connell, 2015; hay bales: Ohara et al., 2015; platforms: Norring et al., 2016), and females exploit the enrichment more than the (heavier) males (Estevez et al., 2002; Ohara et al., 2015). The design of commercially appropriate feeding regimes (e.g. distribution of the dietary ration on the litter surface and wheat and mealworm provision in addition to the dietary ration) to stimulate activity has largely been unsuccessful (Bizeray et al., 2002a,b; Jordan et al., 2011). The provision of oat hulls has, however, been shown to promote dust-bathing and foraging and reduce lameness (Baxter et al., 2018). The provision of slow-growing breeds with an outdoor range has many fundamental welfare benefits over indoor systems, including the encouragement of increased locomotor activity (Nielsen et al., 2003;

Sosnówka-Czajka et al., 2007) and the potential for lowering lameness levels (Zhao et al., 2014; Ipek and Sozcu, 2017).

Selection for fast early growth rate has made broilers susceptible to developing leg disorders as a consequence of significant morphological alterations and a predisposition to inactivity. This, in turn, produces direct welfare problems and confers difficulties in improving housing design. Although the majority of broiler welfare problems have a genetic basis, many, such as lameness, are further exasperated by interactions with a poor environmental management. The provision and maintenance of high-quality litter is very important for general flock health. Contact dermatitis is a skin condition frequently seen in commercial broiler flocks, being particularly prevalent in lame birds since it affects those skin surfaces that have prolonged contact with poor quality (wet) litter.

3 Contact dermatitis

Broilers are particularly susceptible to developing contact dermatitis, due to management practices (particularly litter contamination with faecal ammonia and water) and low activity levels. Contact dermatitis can range in severity from dark skin discolouration and superficial erosions (mild) to pronounced inflammation of subcutaneous tissue and deep necrotic lesions (severe). Most commonly the plantar surface of the feet, including the skin of the central pad and toes, is affected, and this is referred to as footpad dermatitis (FPD) or pododermatitis. Contact dermatitis is also commonly observed to affect the skin of the hock joint and that overlying the sternum, termed 'hock burn' (HB) and breast burn, accordingly.

3.1 Prevalence

A wide variation in the occurrence of moderate to severe FPD has been reported, both between and within studies (e.g. Italy: 0–90%, Meluzzi et al., 2008a,b; USA: 5%, Opengart et al., 2018; Denmark: 13.1%, Lund et al., 2017; Bulgaria: 16%, Dinev et al., 2019; Canada: 28.7%, Hunter et al., 2017; Europe: 37.3%, Bassler et al., 2013; the Netherlands: 38%, de Jong et al., 2012a; UK: 0–48%, Jones et al., 2005; France: 83%, Allain et al., 2009). Although the majority of these studies used a three-point scale, some, for example Allain et al. (2009), used complex scoring systems which can make inter-study comparisons difficult.

Higher prevalence and severity of FPD in European, compared to American, flocks could reflect multiple factors. These include the common European practice of flock thinning (see Section 3.3.1), the type of bedding material (chopped straw litter is frequently used in some parts of Europe while wood-shavings were used in the American study), a more pronounced seasonal (winter) effect in certain parts of Europe, concrete floors in broiler houses (more likely to produce condensation and trap litter moisture compared with the dirt floors typically found in U.S. broiler houses) and the utilisation of a relatively thin layer of bedding, completely replaced between flocks (compared to the deeper build-up and re-use of litter common in the USA) (Opengart et al., 2018).

Although HB develops more slowly than FPD (Skrbic et al., 2015) and is observed less frequently, the two are often associated (e.g. Bassler et al., 2013). As with FPD, a wide variation in HB prevalence has also been reported (e.g. Bulgaria: <1%, Dinev et al., 2019; UK: 0-33%, Haslam et al., 2007; Hepworth et al., 2011; Italy: 3-87%, Melluzzi et al., 2008a,b; Europe: 7.9%, Bassler et al., 2013; France: 59%, Allain et al., 2009), presumably for similar reasons. Although the presence of breast burn and severe HB has been positively correlated (Allain et al., 2009), breast burn is much less frequently seen (e.g. UK: <1%, Haslam et al., 2007; Bulgaria: 5%, Dinev et al., 2019; France: 16%, Allain et al., 2009; Portugal: 18%, Gouveia et al., 2009) and, as a consequence, the condition remains comparatively understudied.

3.2 Qualitative assessment

Many different scoring systems have historically been used to study FPD and HB, ranging from simple binary scores to complex severity scales with as many as ten different categories (e.g. Allain et al., 2009). To enable a meaningful assessment of dermatitis prevalence over time and between systems, it is important to establish a standardised measure. Widely used scoring systems for categorising FPD and HB lesion severity in broilers on farm as part of the 'Welfare Quality (2009) Assessment Protocol for Poultry' utilise five-point severity scales and photographic reference images (Fig. 7). To account for the likelihood that FPD lesion severity is unevenly distributed over commercial broiler units, it is recommended that at least five different sampling locations are employed within each unit and a minimum of 100 birds are scored (de Jong et al., 2012b).

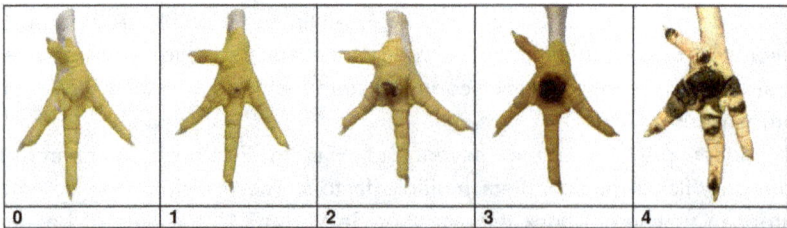

Figure 7 A five-point scoring scale (0-4) for categorising the severity of foot pad dermatitis lesion in broilers (Welfare Quality®, 2009).

3.3 Contact dermatitis risk factors

3.3.1 Age and stocking density

Older broilers are heavier, less active (e.g. Alvino et al., 2009), and consequently experience more skin contact with increasingly faecal-loaded litter. Although contact dermatitis is often reported as increasing with age (FPD: Kyvsgaard et al., 2013; Sarica et al., 2014; HB: Kjaer et al., 2006; Haslam et al., 2007), there are exceptions. FPD lesions have been observed as early as 7 days post-hatch when litter moisture is low (Berk, 2009; Hashimoto et al., 2011). This suggests that young broilers may be more vulnerable to acquiring FPD, with or without exposure to poor litter. There is further evidence for an age-related increase in skin resistance to environmental irritation. Early exposure to high-moisture litter has been shown to increase FPD prevalence and severity (at 14 days); however, no further increase was observed following litter wetting at >56 days of age (Cengiz et al., 2011).

A positive relationship between contact dermatitis and stocking density has often been reported (FPD: Haslam et al., 2007; Bailie et al., 2018b; Ventura et al., 2010; Kyvsgaard et al., 2013; HB: Karaarslan and Nazligul, 2018), associated with high litter moisture and pH (Dozier et al., 2006; Petek et al., 2010). However, stocking density is not a definitive indicator of poor welfare as several studies report no direct impact of density on FPD (Dawkins et al., 2004; Haslam et al., 2006; Meluzzi et al., 2008a; Allain et al., 2009).

Some systems stock at high density and incorporate flock thinning into their production cycle. This involves the removal of a proportion of the flock, usually 1 week prior to the end of production, in order to maintain stocking density (body mass per area) within legal limits. More severe FPD has been reported in birds at final depopulation than in younger thinned-out birds, consistent with the theory that the risk increases with age; however, litter improvement following a reduction in stocking density (post-thinning) may actually enable foot pad health to improve with age (de Jong et al., 2012a).

It has been suggested that placement density (m²/bird) may be a more important factor than finishing density (kg/m²), assuming other environmental factors that impact FPD remain constant. FPD lesions have been observed to form relatively early in life (between 12 and 21 days), correlating with deteriorating litter conditions (including increases in ammonia and moisture), and then to persist over the life of the flock (Opengart et al., 2018).

3.3.2 Sex and body mass

There is plenty of evidence to indicate that male and heavier birds are more at risk of developing HB (Oviedo-Rondón et al., 2009b; Henriksen et al.,

2016; Louton et al., 2018). Indeed body mass at only 2 weeks of age has been identified as a predictor of those flocks most at risk of developing HB prior to slaughter (Hepworth et al., 2010). Although some consider male birds to be more pre-disposed to developing FPD than females due to a heavier body mass (e.g. Gouveia et al., 2009; Sarica et al., 2014; Villarroel et al., 2018), not all studies are in agreement with this. Conflicting observations may reflect a loss of condition (and thus body mass) in broilers with severe FPD, especially if the lesion becomes the site of a secondary infection (e.g. Kyvsgaard et al., 2013).

3.3.3 Production systems

Contact dermatitis lesions are more prevalent/severe in the fast-growing than slower-growing genotypes (FPD: Sarica et al., 2014; Yamak et al., 2016; HB: Haslam et al., 2007; Skrbic et al., 2015), presumably due to different feed conversion rates (and related excreta properties) and activity levels. Unsurprisingly, both FPD prevalence and severity are higher in solid-floor than wire-floor (cage) housing systems (Cengiz et al., 2013; Simsek et al., 2014).

A higher prevalence of FPD has been reported in extensive indoor systems compared to traditional free-range farms (Portugal: Gouveia et al., 2009), and lower HB has been observed in organically reared broilers than in conventional flocks (Broom and Reefmann, 2005; Tuyttens et al., 2008). However, other studies have reported higher FPD scores in flocks with outdoor access than those without (UK: Pagazaurtundua and Warriss, 2006; Turkey: Sarica et al., 2014). The feet of free-range birds will be exposed to ammonia-rich litter in the barns at night, in addition to hard stony ground and pathogens (from wild birds/rodents) while they are out on the range. Broilers with access to a range incorporating both trees and grass have been reported to have better foot condition than those provided with grass alone (Dal Bosco et al., 2014), and this improvement is likely to reflect an increase in range use. Since range conditions, particularly wetness, are an additional risk factor for the pathology (Sans et al., 2014), a well-considered range design, (including the provision of shelter and drainage) should benefit foot health by allowing broilers feet to dry during their time spent outside.

3.3.4 Litter quality

The Welfare Quality Assessment Protocol for broilers describes a five-point scale for scoring litter quality ranging from '0: Completely dry and flaky, moves easily with foot' to '4: Sticks to boots once the cap or compacted crust is broken' (Welfare Quality, 2009). The prevalence and severity of contact dermatitis are often correlated with deterioration in litter quality, particularly when the surface of the litter becomes compacted and the moisture content increases (FPD: Kyvsgaard et al., 2013; de Jong et al., 2014; HB: Allain et al., 2009; Bassler et al., 2013). Wet litter and litter containing a high nitrogen content (from animal waste) are generally more alkaline, that is have a higher pH value (Meluzzi et al., 2008b; Abd El-Wahab et al., 2013). Indeed, improvements in litter quality have been shown to reduce FPD lesion severity in market-age broilers (Cengiz et al., 2011).

Litter quality is related to the type of bedding material used, nutrition, gut health and the optimal management of the housing environment (including heating, ventilation and minimising water spillage). The use of misting systems has been linked with higher levels of FPD (Jones et al., 2005), while the provision of more drinkers per unit area leads to higher litter moisture and a greater prevalence of HB (Jones et al., 2005). Increased water consumption has also been associated with higher FPD (Manning et al., 2007) and HB (Hepworth et al., 2010). Litter quality is presumably reduced directly via increased water spillage and indirectly via increased (and loose) excreta.

Several studies have highlighted that, at least under commercial conditions, litter condition is likely to be more critical than stocking density in the development of FPD (Dawkins et al., 2004; Haslam et al., 2007). Stocking density can contribute towards litter moisture levels; however, with appropriate gut health and effective environmental management, the maintenance of litter quality appears to be possible throughout the production cycle, even under high stocking densities.

3.3.5 Diet

Nutrition, including diet composition and feeding programs, plays a significant role in the aetiology of contact dermatitis due to feed conversion rate, water intake, properties of excreta and the resulting litter quality (see Section 3.3.4). Links between physiological stress, elevated corticosterone and increased water content of faeces are well established (e.g. Puvadolpirod and Thaxton, 2000; Nicol et al., 2009), most likely due to sub-optimal gut function. Jacob et al. (2016a) observed an increase in FPD prevalence between 19 and 29 days old, which correlated with a dietary switch from starter to grower feed; this abrupt dietary transition was hypothesised to reduce gut health and, therefore, litter quality. Presumably the risk of causing stress to the digestive system could easily be reduced by introducing a novel diet gradually, initially presented as a low proportion of the ration and increased over time.

Studies outlining the optimal dietary levels of crude protein, biotin and electrolytes, and the effects of diet composition and nutrient concentration on litter quality and FPD occurrence in broilers, are reviewed in detail elsewhere (e.g. Swiatkiewicz et al., 2017). Dietary constituents vary between countries, and many studies have trialled novel protein sources. FPD is significantly increased in broilers receiving an entirely vegetable-based protein diet compared to those given a diet containing both vegetable- and animal-based proteins

(Nagaraj et al., 2007; Cengiz et al., 2013), and further FPD reductions can be achieved with the addition of corn gluten meal (Eichner et al., 2007). Barley-based diets have been associated with poor litter quality and FPD increase compared to corn-based diets (Cengiz et al., 2017). On the basis of FPD prevalence, rapeseed meal has also been identified as an unsuitable alternative dietary protein to soyabean meal (Abd El-Wahab et al., 2018).

Dietary additives have also been well-researched. The provision of an increased percentage of dietary wheat has been linked with reduced HB (Haslam et al., 2007) and is also seen to benefit leg health (see Section 2.7.4). The inclusion of reduced levels of 2-hydroxy-4-(methylthio)butanoic acid-chelated trace minerals into broiler diets as an alternative to industry levels of inorganic trace minerals has been found to significantly reduce FPD (Zhao et al., 2010; Manangi et al., 2012; Da Costa et al., 2016). Rations supplemented with additional fat (Fuhrmann and Kamphues, 2016), fatty acids (Khosravinia, 2015), vitamin D_3 (Sun et al., 2013), biotin (Abd El-Wahab et al., 2013; Sun et al., 2017), tannic acid (Cengiz et al., 2017) and a directly fed microbe (*Bacillus*) in combination with dietary enzymes (Dersjant-Li et al., 2015) have all been shown to reduce FPD. Dietary enzymes, such as phytase, are beneficial to foot health as they improve nutrient digestibility and reduce litter moisture content (Farhadi et al., 2017).

3.3.6 Lighting

Low-level lighting (0.5-5 lux) delivered with a short scotoperiod (period of darkness) appears to be a risk factor for contact dermatitis (Deep et al., 2013; Bassler et al., 2013; Schwean-Lardner et al., 2013; Senaratna et al., 2016) compared to birds reared under brighter lighting (10-320 lux). This is likely to be associated with an increase in body mass in the low-light birds (Lien et al., 2007; Blatchford et al., 2012; Senaratna et al., 2016). The type of lighting may also be important. The use of LED bulbs has been linked with improved FPD and HB scores when compared with compact fluorescent bulbs (Huth and Archer, 2015).

3.3.7 Egg quality, incubation and early life

At least some susceptibility to contact dermatitis appears to have its foundations in pre-hatch conditions. Lower HB scores have been reported for broilers originating from young parent breeders (particularly those individuals with a low hatching weight) than those originating from older parents (Henriksen et al., 2016).

Appropriate incubation temperatures also appear to be important. Oviedo-Rondón et al. (2009a) observed HB lesions to be more prevalent in 8-week-old commercial broilers hatched from multi-stage incubators (associated with greater temperature fluctuation) than those hatched from single-stage incubators. Fluctuating incubation temperatures were also seen to increase FPD susceptibility via alterations in foot skin structure (Da Costa et al., 2016). Broilers incubated at temperatures (39-40°C) higher than the optimum recommended range may also have a greater risk of developing severe HB in later life (Ipek and Sozcu, 2016).

Raising brooding temperature may help reduce the risk of HB development in heavy chicks. Henriksen et al. (2016) reported that a high brooding temperature (37°C) during the first week post-hatch increased activity, delayed body weight gain and reduced the prevalence of HB at 5 weeks in both high- and low-hatch-weight chicks (compared to high-hatch-weight chicks brooded at 33°C).

3.4 Welfare impact of contact dermatitis

Although there is no evidence for a relationship between contact dermatitis and mortality (de Jong et al., 2012a), skin lesions provide an obvious point of entry for pathogenic microorganisms, which can lead to secondary infections and advanced pathologies such as gangrenous dermatitis and osteomyelitis (Dinev, 2009). In broilers, both FPD and HB are associated with an increased prevalence and severity of *Campylobacter* infection (Bull et al., 2008; Rushton et al., 2009). Hepworth et al. (2011) reported HB to be a useful indicator of flock health since HB was positively associated with the percentage of birds with septicaemia and fever, detected at post-mortem inspection. In addition HB is positively associated with lameness (Sørensen et al., 2000; Haslam et al., 2007); presumably as broilers age and become heavier (and experience deteriorations in leg health), they spend more time lying down, increasing contact time with the litter.

De Jong et al. (2014) observed reductions in production parameters (including body weight) and walking ability (increased GS) in broilers reared on wet litter. As FPD also increased on wet litter the authors hypothesised that birds with deep or infected lesions may have experienced pain and consequently fed and drunk less due to inappetence (de Jong et al., 2014). Sherlock et al. (2012) compared global hepatic gene expression in control birds and in those with experimentally induced FPD and HB lesions. They reported evidence for the inflammatory reaction to impact upon key pathways linked with growth, metabolism and energy utilisation and hypothesised that pain may be the underlying trigger for the up-regulation of genes linked to a pro-inflammatory response and energy metabolism (Sherlock et al., 2012). Further evidence for FPD having a pain component is provided by Hothersall et al. (2016); birds with more severe FPD lesions sat sooner in an LTL test; however, standing ability was improved following NSAID treatment.

3.5 Prevention and control of contact dermatitis

3.5.1 Genetic selection

Since fast-growing genotypes have conclusively been shown to demonstrate greater susceptibility to developing contact dermatitis than the slower-growing genotypes (Sarica et al., 2014; Yamak et al., 2016), greater use of slow-growing strains in extended production systems would have direct and obvious benefits to broiler welfare. HB prevalence also varies according to the standard commercial hybrid used, and this diversity has been attributed to differences in both susceptibility and predisposition to generate wet litter (DEFRA, 2010). Breeding programmes have been shown to offer a direct means by which flock susceptibility to FPD and HB can be selected against and reduced without compromising further improvements in production parameters (Kjaer et al., 2006; Akbas et al., 2009; Ask, 2010; Kapell et al., 2012b), and such selection should continue.

3.5.2 Litter properties

The use of appropriate bedding material in floor-based poultry production systems is necessary to meet several health and welfare requirements. In regard to managing dermatitis deep litter systems should be 'deep', with at least a 7.6 cm depth of litter provided (Shepherd et al., 2017). Bedding material should have high absorbance properties (i.e. have a small particle size) as well as the ability to quickly release moisture (Bilgili et al., 2009; Cengiz et al., 2011). Flocks reared on wood-shavings or sawdust exhibit less FPD than those reared on chopped straw (e.g. Kyvsgaard et al., 2013; Skrbic et al., 2015; Villarroel et al., 2018), rice husks (e.g. Jacob et al., 2016b), grass (e.g. Garcia et al., 2012) or corncob litter (Xavier et al., 2010). Wood-shavings also appear to be the most appropriate litter for managing HB. Lower HB scores were observed at the end of the grow-out period in broilers reared on wood-shavings than paper sludge (Villagrá et al., 2011) or chopped straw (Skrbic et al., 2015). Unchopped straw is unsuitable for use as litter because it has low moisture absorbency and is associated with high FPD (Đukić Stojčić et al., 2016).

Although broiler chickens are commonly housed on sawdust and wood-shavings in some parts of Europe and North America, due to the lack of availability elsewhere, alternative litter sources are required in many parts of the world. Other bedding materials with a potential for maintaining healthy skin include pelleted straw (Berk, 2009; Avdalovic et al., 2017), sand (Simsek et al., 2009) and vermiculite (Yildiz et al., 2014). Similar levels of HB and FPD are reported for conventional wood-shaving litter systems and alternative perforated flooring (netting) systems (Li et al., 2017). The use of shredded paper, although cheap, readily available and highly absorbent, has proved to be unsuccessful in the control of contact dermatitis (Villagrá et al., 2011; Kheravii et al., 2017) since it becomes sodden and fails to release moisture.

In some countries, due to costs and availability, litter (at least the lower layer) is often re-used. Although some studies report a higher prevalence of contact dermatitis in broilers reared on re-used litter than in those reared on fresh (HB: Jacob et al., 2016b; FPD: de Oliveira et al., 2015; Shepherd et al., 2017), other studies report the opposite (Jacob et al., 2016b; Yamak et al., 2016). The ability to successfully re-use sawdust litter appears to be limited to production facilities located in hot dry climates, which exploit effective litter management, ventilation and composting systems.

Litter additives have been shown to be useful in improving litter quality, such as the addition of microorganisms and enzymes to increase the metabolisation of organic components. The application of a biological promoter aimed at wastewater treatment (Micropan®Simplex, Eurovix USA) to chopped straw successfully reduced both litter pH and FPD prevalence compared to a control flock (Đukić Stojčić et al., 2016; Zikic et al., 2017). The addition of sodium bisulphate complex (SBS) to litter has also been shown to be beneficial for foot pad health; this treatment also acidifies the litter, thereby reducing the microbial growth of nitrifying bacteria and NH_3 formation (Toppel et al., 2019).

3.5.3 Managing litter moisture

Litter moisture levels are directly linked to air temperature and RH within the broiler house, which are themselves directly influenced by ambient air temperature and season, the choice of heating system, the ventilation rate, the stocking density and broiler body mass. Deviations in environmental parameters outside of the optimum range recommended by the breeder companies have been identified as a key factor adversely influencing FPD (Jones et al., 2005). A detrimental effect of stocking density on RH has been observed to occur towards the end of the production cycle, indicating that environmental control becomes more difficult at this time (Jones et al., 2005). Presumably the ventilation rate requires to be increased once the birds get larger, less active and cover more of the litter surface, to improve air flow at the litter surface and remove litter moisture. With appropriate environmental management/control and the maintenance of temperature and RH levels within optimum limits, it should be possible to maintain good-quality dry litter throughout the production cycle.

Season has a marked influence upon contact dermatitis, with more lesions observed during colder months (Meluzzi et al., 2008a; Hepworth et al., 2010; Dinev et al., 2019). RH and litter moisture are also usually higher during colder months (e.g. Dawkins et al., 2004; Hermans et al., 2006; Meluzzi et al., 2008a). To save heating costs and maintain shed temperatures, ventilation rates are often reduced at this time. Gas-burning space heaters are a common form of heating in older UK broiler houses (Jones et al., 2005). These systems are located within the sheds themselves and generate a lot of water vapour as a by-product of combustion.

Gut health is linked with litter quality as any increase in faecal water content will be directly transferred to the litter. Causes of diarrhoea (scour) in broilers include coccidiosis, worms, viral or bacterial infection, a diet too high in protein, abrupt dietary alterations or long periods without feed (e.g. caused by feed equipment failures) and biosecurity breaches (Hermans et al., 2006). The use of antibiotics has been shown to improve broiler gut health, indirectly improve litter quality and reduce FPD severity (de Jong et al., 2012a). Global pressure to reduce antibiotic usage in all forms of livestock production has recently sparked much interest in the impact of intestinal microbiota upon broiler productivity (improved feed conversion efficiency) and health (e.g. Stanley et al., 2014; Oviedo-Rondón, 2019). Healthy gut microbiota lowers the risk of enteric diseases due to a reduction in pathogenic load (via competitive exclusion) and improved immune function. A reduction in enteric diseases should, in turn, improve litter quality. The use of probiotics, prebiotics and synbiotics (mixtures of both) to promote beneficial gut microbiota is emerging as a possible alternative to in-feed antibiotics. Results are very encouraging; some studies have even demonstrated prebiotic treatment to be more effective than antibiotics at maintaining broiler intestinal health (e.g. Al-Baadani et al., 2016).

The majority of broiler farms now use nipple and/or cup drinkers as opposed to bell drinkers since they minimise water spillage, and water spillage directly reduces litter quality. Increased water consumption has also been directly associated with FPD (Manning et al., 2007). Daily water consumption is currently monitored in l/bird/day to monitor flock performance and provide an indication of bird health. However, it has been shown that monitoring water consumption as l/m² floor area/day is a good lag (end of crop) indicator of litter quality (Manning et al., 2007). The use of automatic water meters has been linked with a decrease in HB (Hepworth et al., 2010). An ability to monitor and control water intake may, therefore, facilitate better stockmanship and improve litter management.

3.5.4 Environmental enrichment

Although evidence for a positive influence of elevated structures on contact dermatitis is variable and likely to reflect a different uptake in perch use between studies, the provision of straw bales and perches has been reported to reduce contact dermatitis (FPD: Ventura et al., 2010; Hongchao et al., 2013; Ohara et al., 2015; Kiyma et al., 2016; HB: Karaarslan and Nazligul, 2018). Elevated structures encourage better vertical distribution of birds within the available space, decreasing bird density at the floor level and allowing better circulation of air at the litter surface. They also provide a cleaner and dryer surface for the birds to sit and stand on, away from the litter surface.

3.5.5 Monitoring and surveillance systems

Camera systems are being developed to continuously model and monitor the movement and activity of broiler flocks throughout the production cycle to automatically detect and trigger alerts for FPD and HB (Dawkins et al., 2017; Fernandez et al., 2018). One system, in particular, shows great potential to inform improved flock management by providing farmers with an early warning of welfare issues. A camera-based optical flow system (OPTICFLOCK, see Chapter 7) has been shown to have a greater capacity to predict the prevalence of FPD and HB at slaughter (even before external signs become visible) than water consumption, bodyweight or mortality data (Dawkins et al., 2017).

FPD is an important welfare indicator. It is encouraging that government surveillance schemes, whereby producers have to reduce their stocking density or correct management deficiencies (e.g. improve litter quality) if their overall flock score (calculated from the prevalence and severity of FPD lesions at slaughter) exceeds certain predefined trigger limits, have proved to be successful at reducing FPD in Sweden, Denmark (e.g. Kyvsgaard et al., 2013) and the UK (Part et al., 2016). It is hoped that other countries will adopt similar national monitoring schemes and that a widespread uptake of automated monitoring systems will soon provide a means for continuous on-farm surveillance.

4 Conclusion and future trends

Broiler lameness and contact dermatitis remain commonplace within intensively reared commercial flocks, yet it is difficult to infer whether there have been significant improvements or indeed global declines in broiler welfare over recent years. Drawing any definitive conclusions is problematic since data remain sparse and are collected in different ways, multiple management differences exist both within and between countries and the breeds are undergoing constant change (via genetic selection). Leg problems certainly occur more frequently in heavy fast-growing strains due to muscular tissue accumulation occurring at a faster rate than the developing skeleton can support. Although lameness becomes more apparent at higher stocking densities (likely due to forced inactivity), environmental management, rather than stocking density, appears to be critical in the prevention of wet litter and contact dermatitis. Appropriate ventilation should be provided year-round, and both temperature and RH should remain within the recommended range at all times.

Broilers display low activity levels as a consequence of genetic selection for an unnatural body morphology and an excessive body mass; accordingly, they spend the majority of their time sitting in contact with the litter surface. The provision of elevated structures (such as straw bales and platforms with ramp-access), in combination with appropriate lighting, should facilitate rest and encourage activity, ultimately benefiting bone strength and leg and skin health. The provision of slow-growing genotypes with a suitable well-drained range should potentially encourage broilers to perform more activities outside, away from the litter surface of the barn, and thereby limit dermatitis progression.

Both lameness and contact dermatitis pose serious welfare concerns. Lameness physically limits behavioural expression and compromises mobility; birds may experience difficulty in gaining access to feeders and drinkers. Skin lesions provide a gateway for bacteria which can, in turn, trigger joint inflammation. Although evidence for an association between lameness, pain and underlying pathologies remains inconclusive, it is highly likely that some forms of lameness and severe footpad lesions are painful. More research is required to quantify the level of pain associated with specific pathologies, to enable us to differentiate these from other, morphological, gait abnormalities and to better focus breeding programmes (i.e. provide the genetics companies with targeted pressure).

Contact dermatitis and many leg pathologies are heritable and can be selected against. Although the use of genetic selection has had some success in the control of certain pathologies, the persistence of leg and skin conditions in commercial flocks indicates that more needs to be done. A greater uptake, commercially, of the slower-growing breeds would directly benefit broiler welfare. Non-governmental organisations (NGOs) have been, and will continue to be, fundamental in highlighting the requirement for robust genotypes. The Royal Society for Protection of Cruelty to Animals (RSPCA, UK) independently assesses new breeds according to a broiler breed welfare assessment protocol prior to its acceptance for use within the RSPCA-assured scheme; this includes key welfare outcome measures such as growth rate, leg heath and mortality. A number of animal protection organisations have recently set out an NGO 'Broiler Ask' initiative in the EU and the USA. The EU letter can be found at https ://welfarecommitments.com/europeletter/. This scheme directly targets food suppliers and retailers to request that they raise welfare standards for meat chickens including better labelling on meat products, a maximum stocking density of 30 kg/m^2, the adoption of approved higher-welfare breeds and improved lighting by 2026.

Lameness, FPD and HB are generally assessed visually using severity scales, either on farm or at the processing plant. All three measures are used as flock welfare indicators and are crucial for setting (and reviewing) targets to progressively improve welfare within the industry. It is hoped that the development of fully automated video surveillance systems for use on farm will soon enable flock behaviour to be continuously monitored and any anomalies detected to be flagged up to provide producers with an early warning prior to the development of potential welfare issues.

Although standard broiler diets have been developed to deliver optimum nutrition, additional supplements have been observed to confer substantial health benefits. Dietary supplementation with prebiotics, probiotics and synbiotics has been shown to improve the gut uptake of beneficial nutrients and minerals, leading to improvements in the leg health. The addition of dietary enzymes improves nutrient digestibility and has been associated with improvements in both litter quality and footpad health. The uptake of increasingly integrated and automated management systems such as Flockman® (a combined lighting and feed control system for broilers) appears to have numerous welfare benefits. Such systems claim to improve digestion, feed conversion, litter quality and leg health via the automation of lighting programmes and provision of meal feeding to broiler flocks.

More needs to be done to help farmers recognise that the implementation of different feeding practices (e.g. whole wheat feeding, meal feeding and sequential feeding) could have a substantial positive impact upon both broiler welfare and farm economics.

The overarching problem preventing fundamental change within the poultry industry for the improvement of broiler skin and leg health is one of financial cost. The majority of factors that increase bird activity (i.e. appropriate environmental enrichment, lighting regimes, diet and gentotype) also improve skin and leg health; however, the producers (and the genetics companies) hesitate to make these changes due to a real or perceived association of increased activity with lowered production (food conversion) parameters. In fact, it is entirely possible that extra production costs resulting from the implementation of certain management practices designed to reduce lameness and/or improve skin health could be compensated for by other gains. For example, improving leg health and increasing flock uniformity have the potential to confer many economical benefits, including better technical performance (and thus reduced production costs), lowered mortality and reduced carcass condemnation rates. Comprehensive economic cost–benefit analyses may therefore be necessary to bolster both consumer pressure and ethical persuasion for change.

5   Where to look for further information

Further reading

- *Measuring and Auditing Broiler Welfare* by Weeks and Butterworth (2004) provides a good introduction to lameness and contact dermatitis.

- For a useful recent review of the applied ethology of broilers, including space use, stocking density, behaviour and welfare, see *The Behavioural Biology of Chickens* by Nicol (2015).
- *Poultry Feathers and Skin: The Poultry Integument in Health and Welfare*, edited by Olukosi et al. (2019), contains chapters on contact dermatitis and the genetics of contact dermatitis.
- Bradshaw, R. H., Kirkden, R. D. and Broom, D. M. 2002. A review of the aetiology and pathology of leg weakness in broilers in relation to their welfare. *Avian and Poultry Biological Reviews* 13(2), 45-103.
- Dinev, I. 2012. Leg weakness pathology in broiler chickens. *Journal of Poultry Science* 49(2), 63-67.
- Kierończyk, B., Rawski, M., Józefiak, D. and Świątkiewicz, S. 2017. Infectious and non-infectious factors associated with leg disorders in poultry - a review. *Annals of Animal Science* 17(3), 645-69.
- Mayne, R. 2005. A review of the aetiology and possible causative factors of foot pad dermatitis in growing turkeys and broilers. *World's Poultry Science Journal* 61(2), 256-67.
- Pedersen, I. J. and Forkman, B. 2019. Improving leg health in broiler chickens: a systematic review of the effect of environmental enrichment. *Animal Welfare* 28(2), 215-30.
- Swiatkiewicz, S., Arczewska-Wlosek, A. and Jozefiak, D. 2017. The nutrition of poultry as a factor affecting litter quality and foot pad dermatitis - an updated review. *Journal of Animal Physiology and Animal Nutrition (Berlin)* 101(5), e14-20. doi:10.1111/jpn.12630.
- Waldenstedt, L. 2006. Nutritional factors of importance for optimal leg health in broilers: a review. Animal Feed Science and Technology 126(3-4), 291-307.
- Wideman, R. F. 2016. Bacterial chondronecrosis with osteomyelitis and lameness in broilers: a review. *Poultry Science* 95(2), 325-44.

Key Journals

- Animal Welfare
- British Poultry Science
- Journal of Applied Poultry Research
- Poultry Science
- Veterinary Record

Key Conferences

- European Symposium on Poultry Welfare
- UFAW Animal Welfare Conference: Recent Advances in Animal Welfare Science

- WAFL: International Conference on the Assessment of Animal Welfare at Farm and Group Level

6 References

Abd El-Wahab, A., Radko, D. and Kamphues, J. 2013. High dietary levels of biotin and zinc to improve health of foot pads in broilers exposed experimentally to litter with critical moisture content. *Poultry Science* 92(7), 1774–82. doi:10.3382/ps.2013-03054.

Abd El-Wahab, A., Visscher, C. and Kamphues, J. 2018. Impact of different dietary protein sources on performance, litter quality and foot pad dermatitis in broilers. *Journal of Animal and Feed Sciences* 27(2), 148–54. doi:10.22358/jafs/90696/2018.

Abdulla, N. R., Loh, T. C., Akit, H., Sazili, A. Q., Foo, H. L., Kareem, K. Y., Mohamad, R. and Abdul Rahim, R. 2017. Effects of dietary oil sources, calcium and phosphorus levels on growth performance, carcass characteristics and bone quality of broiler chickens. *Journal of Applied Animal Research* 45(1), 423–9. doi:10.1080/09712119.2016.1206903.

Aguado, E., Pascaretti-Grizon, F., Goyenvalle, E., Audran, M. and Chappard, D. 2015. Bone mass and bone quality are altered by hypoactivity in the chicken. *PLoS ONE* 10(1), e0116763. doi:10.1371/journal.pone.0116763.

Akbaş, Y., Yalcin, S., Ozkan, S., Kirkpinar, F., Takma, C., Gevrekçi, Y,, Guller, H. C. and Turkmut, L. 2009. Heritability estimates of tibial dyschondroplasia, valgus-varus, foot-pad dermatitis and hock burn in broiler. *Archiv fur Geflugelkunde* 73(1), 1–6.

Al-Baadani, H. H., Abudabos, A. M., Al-Mufarrej, S. I. and Alzawqari, M. 2016. Effects of dietary inclusion of probiotics, prebiotics and Synbiotics on intestinal histological changes in challenged broiler chickens. *South African Journal of Animal Science* 46(2), 157–65. doi:10.4314/sajas.v46i2.6.

Allain, V., Mirabito, L., Arnould, C., Colas, M., Le Bouquin, S., Lupo, C. and Michel, V. 2009. Skin lesions in broiler chickens measured at the slaughterhouse: relationships between lesions and between their prevalence and rearing factors. *British Poultry Science* 50(4), 407–17. doi:10.1080/00071660903110901.

Al-Rubaye, A. A. K., Ekesi, N. S., Zaki, S., Emami, N. K., Wideman, R. F. and Rhoads, D. D. 2017. Chondronecrosis with osteomyelitis in broilers: further defining a bacterial challenge model using the wire flooring model. *Poultry Science* 96(2), 332–40. doi:10.3382/ps/pew299.

Alvino, G. M., Archer, G. S. and Mench, J. A. 2009. Behavioural time budgets of broiler chickens reared in varying light intensities. *Applied Animal Behaviour Science* 118(1-2), 54–61. doi:10.1016/j.applanim.2009.02.003.

Arnould, C. and Faure, J. M. 2004. Use of pen space and activity of broiler chickens reared at two different densities (vol 84, pg 281, 2003). *Applied Animal Behaviour Science* 87(1-2), 153–70.

Ask, B. 2010. Genetic variation of contact dermatitis in broilers. *Poultry Science* 89(5), 866–75. doi:10.3382/ps.2009-00496.

Avdalovic, V., Vueinic, M., Resanovic, R., Avdalovic, J., Maslic-Strizaks, D. and Vucicevic, M. 2017. Effect of pelleted and chopped wheat straw on the footpad dermatitis in broilers. *Pakistan Journal of Zoology* 49(5), 1639–46. doi:10.17582/journal.pjz/2017.49.5.1639.1646.

Aydin, A. 2017a. Development of an early detection system for lameness of broilers using computer vision. *Computers and Electronics in Agriculture* 136, 140-6. doi:10.1016/j.compag.2017.02.019.

Aydin, A. 2017b. Using 3D vision camera system to automatically assess the level of inactivity in broiler chickens. *Computers and Electronics in Agriculture* 135, 4-10. doi:10.1016/j.compag.2017.01.024.

Aydin, A., Bahr, C. and Berckmans, D. 2015. Automatic classification of measures of lying to assess the lameness of broilers. *Animal Welfare* 24(3), 335-43. doi:10.7120/09627286.24.3.335.

Bailie, C. L. and O'Connell, N. E. 2015. The influence of providing perches and string on activity levels, fearfulness and leg health in commercial broiler chickens. *Animal* 9(4), 660-8. doi:10.1017/S1751731114002821.

Bailie, C. L., Ball, M. E. E. and O'Connell, N. E. 2013. Influence of the provision of natural light and straw bales on activity levels and leg health in commercial broiler chickens. *Animal* 7(4), 618-26. doi:10.1017/S1751731112002108.

Bailie, C. L., Baxter, M. and O'Connell, N. E. 2018a. Exploring perch provision options for commercial broiler chickens. *Applied Animal Behaviour Science* 200, 114-22. doi:10.1016/j.applanim.2017.12.007.

Bailie, C. L., Ijichi, C. and O'Connell, N. E. 2018b. Effects of stocking density and string provision on welfare-related measures in commercial broiler chickens in windowed houses. *Poultry Science* 97(5), 1503-10. doi:10.3382/ps/pey026.

Baracho, M. S., Naas, I. A., Bueno, L. G. F., Nascimento, G. R. and Moura, D. J. 2012. Broiler walking ability and toe asymmetry under harsh rearing conditions. *Revista Brasileira de Ciência Avícola* 14(3), 217-22. doi:10.1590/S1516-635X2012000300009.

Bassler, A. W., Arnould, C., Butterworth, A., Colin, L., De Jong, I. C., Ferrante, V., Ferrari, P., Haslam, S., Wemelsfelder, F. and Blokhuis, H. J. 2013. Potential risk factors associated with contact dermatitis, lameness, negative emotional state, and fear of humans in broiler chicken flocks. *Poultry Science* 92(11), 2811-26. doi:10.3382/ps.2013-03208.

Baxter, M., Bailie, C. L. and O'Connell, N. E. 2018. Evalution of dustbathing substrate and straw bales as environmental enrichments in commercial broiler housing. *Applied Animal Behaviour Science* 200, 78-85. doi:10.1016/j.applanim.2017.11.010.

Bayram, A. and Özkan, S. 2010. Effects of a 16-hour light, 8-hour dark lighting schedule on behavioral traits and performance in male broiler chickens. *Journal of Applied Poultry Research* 19(3), 263-73. doi:10.3382/japr.2009-00026.

Berg, C. and Sanotra, G. S. 2003. Can a modified latency-to-lie test be used to validate gait-scoring results in commercial broiler flocks? *Animal Welfare* 12(4), 655-9.

Berk, J. 2009. Effect of litter type on prevalence and severity of pododermatitis in male broilers. *Berliner Und Munchener Tierarztliche Wochenschrift* 122(7-8), 257-63.

Bilgili, S. F., Hess, J. B., Blake, J. P., Macklin, K. S., Saenmahayak, B. and Sibley, J. L. 2009. Influence of bedding material on footpad dermatitis in broiler chickens. *Journal of Applied Poultry Research* 18(3), 583-9. doi:10.3382/japr.2009-00023.

Birgul, O. B., Mutaf, S. and Alkan, S. 2012. Effects of different angled perches on leg disorders in broilers. *Archiv fur Geflugelkunde* 76(1), 44-8.

Bizeray, D., Estevez, I., Leterrier, C. and Faure, J. M. 2002a. Effects of increasing environmental complexity on the physical activity of broiler chickens. *Applied Animal Behaviour Science* 79(1), 27-41. doi:10.1016/S0168-1591(02)00083-7.

Bizeray, D., Estevez, I., Leterrier, C. and Faure, J. M. 2002b. Influence of increased environmental complexity on leg condition, performance, and level of fearfulness in broilers. *Poultry Science* 81(6), 767–73. doi:10.1093/ps/81.6.767.

Blatchford, R. A., Klasing, K. C., Shivaprasad, H. L., Wakenell, P. S., Archer, G. S. and Mench, J. A. 2009. The effect of light intensity on the behavior, eye and leg health, and immune function of broiler chickens. *Poultry Science* 88(1), 20–8. doi:10.3382/ps.2008-00177.

Blatchford, R. A., Archer, G. S. and Mench, J. A. 2012. Contrast in light intensity, rather than day length, influences the behavior and health of broiler chickens. *Poultry Science* 91(8), 1768–74. doi:10.3382/ps.2011-02051.

Bokkers, E. A. M. and Koene, P. 2003. Behaviour of fast- and slow growing broilers to 12 weeks of age and the physical consequences. *Applied Animal Behaviour Science* 81(1), 59–72. doi:10.1016/S0168-1591(02)00251-4.

Bokkers, E. A. M. and Koene, P. 2004. Motivation and ability to walk for a food reward in fast- and slow-growing broilers to 12 weeks of age. *Behavioural Processes* 67(2), 121–30. doi:10.1016/j.beproc.2004.03.015.

Bokkers, E. A. M., Zimmerman, P. H., Rodenburg, T. B. and Koene, P. 2007. Walking behaviour of heavy and light broilers in an operant runway test with varying durations of feed deprivation and feed access. *Applied Animal Behaviour Science* 108(1-2), 129–42. doi:10.1016/j.applanim.2006.10.011.

Bradbury, E. J., Wilkinson, S. J., Cronin, G. M., Thomson, P. C., Bedford, M. R. and Cowieson, A. J. 2014. Nutritional geometry of calcium and phosphorus nutrition in broiler chicks. Growth performance, skeletal health and intake arrays. *Animal* 8(7), 1071–9. doi:10.1017/S1751731114001037.

Bradbury, E. J., Wilkinson, S. J., Cronin, G. M., Thomson, P., Walk, C. L. and Cowieson, A. J. 2017. Evaluation of the effect of a highly soluble calcium source in broiler diets supplemented with phytase on performance, nutrient digestibility, foot ash, mobility and leg weakness. *Animal Production Science* 57(10), 2016–26. doi:10.1071/AN16142.

Bradshaw, R. H., Kirkden, R. D. and Broom, D. M. 2002. A review of the aetiology and pathology of leg weakness in broilers in relation to welfare. *Avian and Poultry Biology Reviews* 13(2), 45–103. doi:10.3184/147020602783698421.

Brickett, K. E., Dahiya, J. P., Classen, H. L., Annett, C. B. and Gomis, S. 2007. The impact of nutrient density, feed form, and photoperiod on the walking ability and skeletal quality of broiler chickens. *Poultry Science* 86(10), 2117–25. doi:10.1093/ps/86.10.2117.

Broom, D. M. and Reefmann, N. 2005. Chicken welfare as indicated by lesions on carcases in supermarkets. *British Poultry Science* 46(4), 407–14. doi:10.1080/00071660500181149.

Buijs, S., Keeling, L., Rettenbacher, S., Van Poucke, E. and Tuyttens, F. A. M. 2009. Stocking density effects on broiler welfare: identifying sensitive ranges for different indicators. *Poultry Science* 88(8), 1536–43. doi:10.3382/ps.2009-00007.

Buijs, S., Keeling, L. J., Vangestel, C., Baert, J. and Tuyttens, F. A. M. 2011. Neighbourhood analysis as an indicator of spatial requirements of broiler chickens. *Applied Animal Behaviour Science* 129(2-4), 111–20. doi:10.1016/j.applanim.2010.11.017.

Buijs, S., Van Poucke, E., Van Dongen, S., Lens, L., Baert, J. and Tuyttens, F. A. 2012. The influence of stocking density on broiler chicken bone quality and fluctuating asymmetry. *Poultry Science* 91(8), 1759–67. doi:10.3382/ps.2011-01859.

Bull, S. A., Thomas, A., Humphrey, T., Ellis-Iversen, J., Cook, A. J., Lovell, R. and Jorgensen, F. 2008. Flock health indicators and *Campylobacter* spp. in commercial housed broilers reared in Great Britain. *Applied and Environmental Microbiology* 74(17), 5408-13. doi:10.1128/AEM.00462-08.

Butterworth, A. and Haslam, S. M. 2009. A lameness control strategy for broiler fowls. Welfare Quality Reports No. 13. Cardiff School of City and Regional Planning, Cardiff, UK. Available at: http://www.welfarequality.net/media/1119/wqr13.pdf.

Butterworth, A., Weeks, C. A., Crea, P. R. and Kestin, S. C. 2002. Dehydration and lameness in a broiler flock. *Animal Welfare* 11(1), 89-94.

Caplen, G., Hothersall, B., Murrell, J. C., Nicol, C. J., Waterman-Pearson, A. E., Weeks, C. A. and Colborne, G. R. 2012. Kinematic analysis quantifies gait abnormalities associated with lameness in broiler chickens and identifies evolutionary gait differences. *PLoS ONE* 7(7), e40800. doi:10.1371/journal.pone.0040800.

Caplen, G., Colborne, G. R., Hothersall, B., Nicol, C. J., Waterman-Pearson, A. E., Weeks, C. A. and Murrell, J. C. 2013a. Lame broiler chickens respond to non-steroidal anti-inflammatory drugs with objective changes in gait function: a controlled clinical trial. *Veterinary Journal* 196(3), 477-82. doi:10.1016/j.tvjl.2012.12.007.

Caplen, G., Baker, L., Hothersall, B., McKeegan, D. E. F., Sandilands, V., Sparks, N. H., Waterman-Pearson, A. E. and Murrell, J. C. 2013b. Thermal nociception as a measure of non-steroidal anti-inflammatory drug effectiveness in broiler chickens with articular pain. *Veterinary Journal* 198(3), 616-9. doi:10.1016/j.tvjl.2013.09.013.

Caplen, G., Hothersall, B., Nicol, C. J., Parker, R. M. A., Waterman-Pearson, A. E., Weeks, C. and Murrell, J. 2014. Lameness is consistently better at predicting broiler chicken performance in mobility tests than other broiler characteristics. *Animal Welfare* 23(2), 179-87. doi:10.7120/09627286.23.2.179.

Castellini, C., Mugnai, C., Moscati, L., Mattioli, S., Amato, M. G., Cartoni Mancinelli, A. and Dal Bosco, A. 2016. Adaptation to organic rearing system of eight different chicken genotypes: behaviour, welfare and performance. *Italian Journal of Animal Science* 15(1), 37-46. doi:10.1080/1828051X.2015.1131893.

Cengiz, Ö, Hess, J. B. and Bilgili, S. F. 2011. Effect of bedding type and transient wetness on footpad dermatitis in broiler chickens. *Journal of Applied Poultry Research* 20(4), 554-60. doi:10.3382/japr.2011-00368.

Cengiz, O., Hess, J. B. and Bilgili, S. F. 2013. Effect of protein source on the development of footpad dermatitis in broiler chickens reared on different flooring types. *Archiv fur Geflugelkunde* 77(3), 166-70.

Cengiz, Ö, Koksal, B. H., Tatli, O., Sevim, Ö, Ahsan, U., Bilgili, S. F. and Gökhan Önol, A. 2017. Effect of dietary tannic acid supplementation in corn- or barley-based diets on growth performance, intestinal viscosity, litter quality, and incidence and severity of footpad dermatitis in broiler chickens. *Livestock Science* 202, 52-7. doi:10.1016/j. livsci.2017.05.016.

Charles, J. F., Ermann, J. and Aliprantis, A. O. 2015. The intestinal microbiome and skeletal fitness: connecting bugs and bones. *Clinical Immunology* 159(2), 163-9. doi:10.1016/j.clim.2015.03.019.

Classen, H. L. 2004. Day length affects performance, health and condemnations in broiler chickens. *Proceedings of the 16th Australian Poultry Science Symposium*, Sydney, New South Wales, Australia, 9-11 February 2004, pp. 11215. Available at: https:// sydney.edu.au/vetscience/apss/documents/2004/APSS2004-classen-pp112-115. pdf.

Collins, L. M. 2008. Non-intrusive tracking of commercial broiler chickens *in situ* at different stocking densities. *Applied Animal Behaviour Science* 112(1-2), 94-105. doi:10.1016/j.applanim.2007.08.009.

Corr, S. A., Gentle, M. J., McCorquodale, C. C. and Bennett, D. 2003a. The effect of morphology on walking ability in the modern broiler: a gait analysis study. *Animal Welfare* 12(2), 159-71.

Corr, S. A., Gentle, M. J., McCorquodale, C. C. and Bennett, D. 2003b. The effect of morphology on the musculoskeletal system of the modern broiler. *Animal Welfare* 12(2), 145-57.

Corr, S. A., Maxwell, M., Gentle, M. J. and Bennett, D. 2003c. Preliminary study of joint disease in poultry by the analysis of synovial fluid. *Veterinary Record* 152(18), 549-54. doi:10.1136/vr.152.18.549.

Corr, S. A., McCorquodale, C., McDonald, J., Gentle, M. and McGovern, R. 2007. A force plate study of avian gait. *Journal of Biomechanics* 40(9), 2037-43. doi:10.1016/j.jbiomech.2006.09.014.

Coto, C., Yan, F., Cerrate, S., Wang, Z., Sacakli, P., Halley, J. T., Wiernusz, C. J., Martinez, A. and Waldroup, P. W. 2008. Effects of dietary levels of calcium and nonphytate P in broiler starter diets on live performance, bone development, and growth plate conditions in male broilers fed a corn-based diet. *International Journal of Poultry Science* 7(7), 638-45. doi:10.3923/ijps.2008.638.645.

Da Costa, M. J., Oviedo-Rondon, E. O., Wineland, M. J., Claassen, K. and Osborne, J. 2016. Effects of incubation temperatures and trace mineral sources on chicken live performance and footpad skin development. *Poultry Science* 95(4), 749-59. doi:10.3382/ps/pev446.

Dal Bosco, A. D., Mugnai, C., Amato, M. G., Piottoli, L., Cartoni, A. and Castellini, C. 2014. Effect of slaughtering age in different commercial chicken genotypes reared according to the organic system: 1. Welfare, carcass and meat traits. *Italian Journal of Animal Science* 13(2), 3308. doi:10.4081/ijas.2014.3308.

Danbury, T. C., Weeks, C. A., Chambers, J. P., Waterman-Pearson, A. E. and Kestin, S. C. 2000. Self-selection of the analgesic drug carprofen by lame broiler chickens. *Veterinary Record* 146(11), 307-11. doi:10.1136/vr.146.11.307.

Dantzer, R. 2009. Cytokine, sickness behavior, and depression. *Immunology and Allergy Clinics of North America* 29(2), 247-64. doi:10.1016/j.iac.2009.02.002.

Das, H. and Lacin, E. 2014. The effect of different photoperiods and stocking densities on fattening performance, carcass and some stress parameters in broilers. *Israel Journal of Veterinary Medicine* 69(4), 211 20.

Dawkins, M. S., Donnelly, C. A. and Jones, T. A. 2004. Chicken welfare is influenced more by housing conditions than by stocking density. *Nature* 427(6972), 342-4. doi:10.1038/nature02226.

Dawkins, M. S., Lee, H. J., Waitt, C. D. and Roberts, S. J. 2009. Optical flow patterns in broiler chicken flocks as automated measures of behaviour and gait. *Applied Animal Behaviour Science* 119(3-4), 203-9. doi:10.1016/j.applanim.2009.04.009.

Dawkins, M. S., Cain, R., Merelie, K. and Roberts, S. J. 2013. In search of the behavioural correlates of optical flow patterns in the automated assessment of broiler chicken welfare. *Applied Animal Behaviour Science* 145(1-2), 44-50. doi:10.1016/j.applanim.2013.02.001.

Dawkins, M. S., Roberts, S. J., Cain, R. J., Nickson, T. and Donnelly, C. A. 2017. Early warning of footpad dermatitis and hockburn in broiler chicken flocks using optical flow, bodyweight and water consumption. *Veterinary Record* 180(20), 499. doi:10.1136/vr.104066.

de Jong, I. C. and Gunnink, H. 2019. Effects of a commercial broiler enrichment programme with or without natural light on behaviour and other welfare indicators. *Animal* 13(2), 384–91. doi:10.1017/S1751731118001805.

de Jong, I. C., van Ham, J., Gunnink, H., Hindle, V. A. and Lourens, A. 2012a. Footpad dermatitis in Dutch broiler flocks: prevalence and factors of influence. *Poultry Science* 91(7), 1569–74. doi:10.3382/ps.2012-02156.

de Jong, I. C., van Harn, J., Gunnink, H., Lourens, A. and van Riel, J. W. 2012b. Measuring foot-pad lesions in commercial broiler houses. Some aspects of methodology. *Animal Welfare* 21(3), 325–30. doi:10.7120/09627286.21.3.325.

de Jong, I. C., Gunnink, H. and van Harn, J. 2014. Wet litter not only induces footpad dermatitis but also reduces overall welfare, technical performance, and carcass yield in broiler chickens. *Journal of Applied Poultry Research* 23(1), 51–8. doi:10.3382/japr.2013-00803.

de Oliveira, M. C., Goncalves, B. N., Padua, G. T., da Silva, V. G., da Silva, D. V. and Freitas, A. I. M. 2015. Treatment of poultry litter does not improve performance or carcass lesions in broilers. *Revista Colombiana De Ciencias Pecuarias* 28(4), 331–8.

Deep, A., Raginski, C., Schwean-Lardner, K., Fancher, B. I. and Classen, H. L. 2013. Minimum light intensity threshold to prevent negative effects on broiler production and welfare. *British Poultry Science* 54(6), 686–94. doi:10.1080/00071668.2013.847526.

Department of the Environment Food and Rural Affairs. 2010. Foot pad dermatitis & hock burn in broilers: risk factors, aetiology & welfare consequences. London, UK. Available at: http://sciencesearch.defra.gov.uk/Document.aspx?Document=AW1137SID5FinalReport.pdf (accessed on 16 June 2019).

Dersjant-Li, Y., van de Belt, K., van der Klis, J. D., Kettunen, H., Rinttila, T. and Awati, A. 2015. Effect of multi-enzymes in combination with a direct-fed microbial on performance and welfare parameters in broilers under commercial production settings. *Journal of Applied Poultry Research* 24(1), 80–90. doi:10.3382/japr/pfv003.

Dinev, I. 2009. Clinical and morphological investigations on the prevalence of lameness associated with femoral head necrosis in broilers. *British Poultry Science* 50(3), 284–90. doi:10.1080/00071660902942783.

Dinev, I. and Kanakov, D. 2011. Deep pectoral myopathy: prevalence in 7 weeks old broiler chickens in Bulgaria. *Revue De Medecine Veterinaire* 162(6), 279–83.

Dinev, I., Denev, S. A. and Edens, F. W. 2012. Comparative clinical and morphological studies on the incidence of tibial dyschondroplasia as a cause of lameness in three commercial lines of broiler chickens. *Journal of Applied Poultry Research* 21(3), 637–44. doi:10.3382/japr.2010-00303.

Dinev, I., Denev, S., Vashin, I., Kanakov, D. and Rusenova, N. 2019. Pathomorphological investigations on the prevalence of contact dermatitis lesions in broiler chickens. *Journal of Applied Animal Research* 47(1), 129–34. doi:10.1080/09712119.2019.1584105.

Dozier, W. A., Thaxton, J. P., Purswell, J. L., Olanrewaju, H. A., Branton, S. L. and Roush, W. B. 2006. Stocking density effects on male broilers grown to 1.8 kilograms of body weight. *Poultry Science* 85(2), 344–51. doi:10.1093/ps/85.2.344.

Duff, S. R. I. and Thorp, B. H. 1985. Abnormal angulation/torsion of the pelvic appendicular skeleton in broiler fowl: morphological and radiological findings. *Research in Veterinary Science* 39(3), 313–9. doi:10.1016/S0034-5288(18)31720-X.

Đukić Stojčić, M., Bjedov, S., Zikic, D., Peric, L. and Milosevic, N. 2016. Effect of straw size and microbial amendment of litter on certain litter quality parameters, ammonia emission, and footpad dermatitis in broilers. *Archives Animal Breeding* 59(1), 131-7. doi:10.5194/aab-59-131-2016.

Eichner, G., Vieira, S. L., Torres, C. A., Coneglian, J. L. B., Freitas, D. M. and Oyarzabal, O. A. 2007. Litter moisture and footpad dermatitis as affected by diets formulated on an all-vegetable basis or having the inclusion of poultry by-product. *Journal of Applied Poultry Research* 16(3), 344-50. doi:10.1093/japr/16.3.344.

Eriksson, M., Waldenstedt, L., Elwinger, K., Engstrom, B. and Fossum, O. 2010. Behaviour, production and health of organically reared fast-growing broilers fed low crude protein diets including different amino acid contents at start. *Acta Agriculturae Scandinavica, Section A – Animal Science* 60(2), 112-24. doi:10.1080/09064702.2010.502243.

Estevez, I. 2007. Density allowances for broilers: where to set the limits? *Poultry Science* 86(6), 1265-72. doi:10.1093/ps/86.6.1265.

Estevez, I., Tablante, N., Pettit-Riley, R. L. and Carr, L. 2002. Use of cool perches by broiler chickens. *Poultry Science* 81(1), 62-9. doi:10.1093/ps/81.1.62.

FAO. 2010. Poultry welfare in developing countries – welfare issues in commercial broiler production, pp. 117-18. Available at: http://www.fao.org/3/a-al723e.pdf.

Farhadi, D., Karimi, A., Sadeghi, G., Rostamzadeh, J. and Bedford, M. R. 2017. Effects of a high dose of microbial phytase and myo-inositol supplementation on growth performance, tibia mineralization, nutrient digestibility, litter moisture content, and foot problems in broiler chickens fed phosphorus-deficient diets. *Poultry Science* 96(10), 3664-75. doi:10.3382/ps/pex186.

Fernandes, B. C. D., Martins, M. R. F. B., Mendes, A. A., Paz, I., Komiyama, C. M., Milbradt, E. L. and Martins, B. B. 2012. Locomotion problems of broiler chickens and its relationship with the gait score. *Revista Brasileira de Zootecnia – Brazilian Journal of Animal Science* 41(8), 1951-5. doi:10.1590/S1516-35982012000800021.

Fernandez, A. P., Norton, T., Tullo, E., van Hertem, T., Youssef, A., Exadaktylos, V., Vranken, E., Guarino, M. and Berckmans, D. 2018. Real-time monitoring of broiler flock's welfare status using camera-based technology. *Biosystems Engineering* 173, 103-14. doi:10.1016/j.biosystemseng.2018.05.008.

Fidan, E. D., Nazligul, A., Turkyilmaz, M. K., Aypak, S. Ü., Kilimci, F. S., Karaarslan, S. and Kaya, M. 2017. Effect of photoperiod length and light intensity on some welfare criteria, carcass, and meat quality characteristics in broilers. *Revista Brasileira de Zootecnia – Brazilian Journal of Animal Science* 46(3), 202-10. doi:10.1590/s1806-92902017000300004.

Fleming, R. H. 2008. Nutritional factors affecting poultry bone health. *Proceedings of the Nutrition Society* 67(2), 177-83. doi:10.1017/S0029665108007015.

Fuhrmann, R. and Kamphues, J. 2016. Effects of fat content and source as well as of calcium and potassium content in the diet on fat excretion and saponification, litter quality and foot pad health in broilers. *European Poultry Science* 80, 118. doi:10.1399/eps.2016.118.

Garcia, R., Almeida Paz, I., Caldara, F., Nääs, I., Bueno, L., Freitas, L., Graciano, J. and Sim, S. 2012. Litter materials and the incidence of carcass lesions in broilers chickens. *Revista Brasileira de Ciência Avícola* 14(1), 27-32. doi:10.1590/S1516-635X2012000100005.

Garner, J. P., Falcone, C., Wakenell, P., Martin, M. and Mench, J. A. 2002. Reliability and validity of a modified gait scoring system and its use in assessing tibial dyschondroplasia in broilers. *British Poultry Science* 43(3), 355-63. doi:10.1080/00071660120103620.

Gentle, M. J., Tilston, V. and McKeegan, D. E. F. 2001. Mechanothermal nociceptors in the scaly skin of the chicken leg. *Neuroscience* 106(3), 643-52. doi:10.1016/s0306-4522(01)00318-9.

Gocsik, É., Silvera, A. M., Hansson, H., Saatkamp, H. W. and Blokhuis, H. J. 2017. Exploring the economic potential of reducing broiler lameness. *British Poultry Science* 58(4), 337-47. doi:10.1080/00071668.2017.1304530.

Gouveia, K. G., Vaz-Pires, P. and da Costa, P. M. 2009. Welfare assessment of broilers through examination of haematomas, foot-pad dermatitis, scratches and breast blisters at processing. *Animal Welfare* 18(1), 43-8.

Granquist, E. G., Vasdal, G., de Jong, I. C. and Moe, R. O. 2019. Lameness and its relationship with health and production measures in broiler chickens. *Animal* 13(10), 2365-72. doi:10.1017/S1751731119000466.

Groves, P. J. and Muir, W. I. 2014. A meta-analysis of experiments linking incubation conditions with subsequent leg weakness in broiler chickens. *PLoS ONE* 9(7), e102682. doi:10.1371/journal.pone.0102682.

Groves, P. J. and Muir, W. I. 2017. Earlier hatching time predisposes Cobb broiler chickens to tibial dyschondroplasia. *Animal* 11(1), 112-20. doi:10.1017/S1751731116001105.

Guo, R. X., Li, Z. Y., Zhou, X. H., Huang, C. Y., Hu, Y. C., Geng, S., Chen, X., Li, Q., Pan, Z. and Jiao, X. 2019. Induction of arthritis in chickens by infection with novel virulent *Salmonella pullorum* strains. *Veterinary Microbiology* 228, 165-72. doi:10.1016/j.vetmic.2018.11.032.

Hashimoto, S., Yamazaki, K., Obi, T. and Takase, K. 2011. Footpad dermatitis in broiler chickens in Japan. *Journal of Veterinary Medical Science* 73(3), 293-7. doi:10.1292/jvms.10-0329.

Hashimoto, S., Yamazaki, K., Obi, T. and Takase, K. 2013. Relationship between severity of footpad dermatitis and carcass performance in broiler chickens. *Journal of Veterinary Medical Science* 75(11), 1547-9. doi:10.1292/jvms.13-0031.

Haslam, S. M., Brown, S. N., Wilkins, L. J., Kestin, S. C., Warriss, P. D. and Nicol, C. J. 2006. Preliminary study to examine the utility of using foot burn or hock burn to assess aspects of housing conditions for broiler chicken. *British Poultry Science* 47(1), 13-8. doi:10.1080/00071660500475046.

Haslam, S. M., Knowles, T. G., Brown, S. N., Wilkins, L. J., Kestin, S. C., Warriss, P. D. and Nicol, C. J. 2007. Factors affecting the prevalence of foot pad dermatitis, hock burn and breast burn in broiler chicken. *British Poultry Science* 48(3), 264-75. doi:10.1080/00071660701371341.

Henriksen, S., Bilde, T. and Riber, A. B. 2016. Effects of post-hatch brooding temperature on broiler behavior, welfare, and growth. *Poultry Science* 95(10), 2235-43. doi:10.3382/ps/pew224.

Hepworth, P. J., Nefedov, A. V., Muchnik, I. B. and Morgan, K. L. 2010. Early warning indicators for hock burn in broiler flocks. *Avian Pathology* 39(5), 405-9. doi:10.1080/03079457.2010.510500.

Hepworth, P. J., Nefedov, A. V., Muchnik, I. B. and Morgan, K. L. 2011. Hock burn: an indicator of broiler flock health. *Veterinary Record* 168(11), 303. doi:10.1136/vr.c6897.

Hermans, P. G., Fradkin, D., Muchnik, I. B. and Morgan, K. L. 2006. Prevalence of wet litter and the associated risk factors in broiler flocks in the United Kingdom. *Veterinary Record* 158(18), 615-22. doi:10.1136/vr.158.18.615.

Hester, P. Y., Enneking, S. A., Haley, B. K., Cheng, H. W., Einstein, M. E. and Rubin, D. A. 2013. The effect of perch availability during pullet rearing and egg laying on musculoskeletal health of caged White Leghorn hens. *Poultry Science* 92(8), 1972–80. doi:10.3382/ps.2013-03008.

Hongchao, J., Jiang, Y., Song, Z., Zhao, J., Wang, X. and Lin, H. 2013. Effect of perch type and stocking density on the behaviour and growth of broilers. *Animal Production Science* 54(7), 930–41. doi:10.1071/AN13184.

Hothersall, B., Caplen, G., Parker, R. M. A., Nicol, C. J., Waterman-Pearson, A. E., Weeks, C. A. and Murrell, J. C. 2014. Thermal nociceptive threshold testing detects altered sensory processing in broiler chickens with spontaneous lameness. *PLoS ONE* 9(5), e97883. doi:10.1371/journal.pone.0097883.

Hothersall, B., Caplen, G., Parker, R. M. A., Nicol, C. J., Waterman-Pearson, A. E., Weeks, C. and Murrell, J. 2016. Effects of carprofen, meloxicam and butorphanol on broiler chickens' performance in mobility tests. *Animal Welfare* 25(1), 55–67. doi:10.7120/09627286.25.1.055.

Hunter, J. M., Anders, S. A., Crowe, T., Korver, D. R. and Bench, C. J. 2017. Practical assessment and management of foot pad dermatitis in commercial broiler chickens: a field study. *Journal of Applied Poultry Research* 26(4), 593–604. doi:10.3382/japr/pfx019.

Huth, J. C. and Archer, G. S. 2015. Comparison of two LED light bulbs to a dimmable CFL and their effects on broiler chicken growth, stress, and fear. *Poultry Science* 94(9), 2027–36. doi:10.3382/ps/pev215.

Ipek, A. and Sozcu, A. 2016. The effects of eggshell temperature fluctuations during incubation on welfare status and gait score of broilers. *Poultry Science* 95(6), 1296–303. doi:10.3382/ps/pew056.

Ipek, A. and Sozcu, A. 2017. The effects of access to pasture on growth performance, behavioural patterns, some blood parameters and carcass yield of a slow-growing broiler genotype. *Journal of Applied Animal Research* 45(1), 464–9. doi:10.1080/09712119.2016.1214136.

Jacob, F. G., Baracho, M. S., Naas, I. A., Salgado, D. A. and Souza, R. 2016a. Incidence of pododermatitis in broiler reared under two types of environment. *Revista Brasileira de Ciência Avícola* 18(2), 247–54. doi:10.1590/1806-9061-2015-0047.

Jacob, F. G., Baracho, M. S., Naas, I. A., Lima, N. S. D., Salgado, D. D. and Souza, R. 2016b. Risk of incidence of hock burn and pododermatitis in broilers reared under commercial conditions. *Revista Brasileira de Ciência Avícola* 18(3), 357–62. doi:10.1590/1806-9061-2015-0183.

James, C., Asher, L., Herborn, K. and Wiseman, J. 2018. The effect of supplementary ultraviolet wavelengths on broiler chicken welfare indicators. *Applied Animal Behaviour Science* 209, 55–64. doi:10.1016/j.applanim.2018.10.002.

Jones, T. A., Donnelly, C. A. and Dawkins, M. S. 2005. Environmental and management factors affecting the welfare of chickens on commercial farms in the United Kingdom and Denmark stocked at five densities. *Poultry Science* 84(8), 1155–65. doi:10.1093/ps/84.8.1155.

Jordan, D., Stuhec, I. and Bessei, W. 2011. Effect of whole wheat and feed pellets distribution in the litter on broilers' activity and performance. *Archiv fur Geflugelkunde* 75(2), 98–103.

Julian, R. J. 1998. Rapid growth problems: ascites and skeletal deformities in broilers. *Poultry Science* 77(12), 1773–80. doi:10.1093/ps/77.12.1773.

Kapell, D. N., Hill, W. G., Neeteson, A. M., McAdam, J., Koerhuis, A. N. M. and Avendaño, S. 2012a. Twenty-five years of selection for improved leg health in purebred broiler lines and underlying genetic parameters. *Poultry Science* 91(12), 3032–43. doi:10.3382/ps.2012-02578.

Kapell, D. N., Hill, W. G., Neeteson, A. M., McAdam, J., Koerhuis, A. N. M. and Avendaño, S. 2012b. Genetic parameters of foot-pad dermatitis and body weight in purebred broiler lines in 2 contrasting environments. *Poultry Science* 91(3), 565–74. doi:10.3382/ps.2011-01934.

Karaarslan, S. and Nazligul, A. 2018. Effects of lighting, stocking density, and access to perches on leg health variables as welfare indicators in broiler chickens. *Livestock Science* 218, 31–6. doi:10.1016/j.livsci.2018.10.008.

Kaukonen, E., Norring, M. and Valros, A. 2017. Perches and elevated platforms in commercial broiler farms: use and effect on walking ability, incidence of tibial dyschondroplasia and bone mineral content. *Animal* 11(5), 864–71. doi:10.1017/S1751731116002160.

Kestin, S. C., Knowles, T. G., Tinch, A. E. and Gregory, N. G. 1992. Prevalence of leg weakness in broiler-chickens and its relationship with genotype. *Veterinary Record* 131(9), 190–4. doi:10.1136/vr.131.9.190.

Kestin, S. C., Gordon, S., Su, G. and Sorensen, P. 2001. Relationships in broiler chickens between lameness, liveweight, growth rate and age. *Veterinary Record* 148(7), 195–7. doi:10.1136/vr.148.7.195.

Kheravii, S. K., Swick, R. A., Choct, M. and Wu, S. B. 2017. Potential of pelleted wheat straw as an alternative bedding material for broilers. *Poultry Science* 96(6), 1641–7. doi:10.3382/ps/pew473.

Khosravinia, H. 2015. Effect of dietary supplementation of medium-chain fatty acids on growth performance and prevalence of carcass defects in broiler chickens raised in different stocking densities. *Journal of Applied Poultry Research* 24(1), 1–9. doi:10.3382/japr/pfu001.

Kittelsen, K. E., David, B., Moe, R. O., Poulsen, H. D., Young, J. F. and Granquist, E. G. 2017. Associations among gait score, production data, abattoir registrations, and postmortem tibia measurements in broiler chickens. *Poultry Science* 96(5), 1033–40. doi:10.3382/ps/pew433.

Kiyma, Z., Kucukyilmaz, K. and Orojpour, A. 2016. Effects of perch availability on performance, carcass characteristics, and footpad lesions in broilers. *Archives Animal Breeding* 59(1), 19–25. doi:10.5194/aab-59-19-2016.

Kjaer, J. B., Su, G., Nielsen, B. L. and Sorensen, P. 2006. Foot pad dermatitis and hock burn in broiler chickens and degree of inheritance. *Poultry Science* 85(8), 1342–8. doi:10.1093/ps/85.8.1342.

Knowles, T. G., Kestin, S. C., Haslam, S. M., Brown, S. N., Green, L. E., Butterworth, A., Pope, S. J., Pfeiffer, D. and Nicol, C. J. 2008. Leg disorders in broiler chickens: prevalence, risk factors and prevention. *PLoS ONE* 3(2), e1545. doi:10.1371/journal.pone.0001545.

Kwiatkowska, K., Winiarska-Mieczan, A. and Kwiecien, M. 2017. Feed additives regulating calcium homeostasis in the bones of poultry – a review. *Annals of Animal Science* 17(2), 303–16. doi:10.1515/aoas-2016-0031.

Kyvsgaard, N. C., Jensen, H. B., Ambrosen, T. and Toft, N. 2013. Temporal changes and risk factors for foot-pad dermatitis in Danish broilers. *Poultry Science* 92(1), 26–32. doi:10.3382/ps.2012-02433.

Leterrier, C., Vallee, C., Constantin, P., Chagneau, A. M., Lessire, M., Lescoat, P., Berri, C., Baéza, E., Bizeray, D. and Bouvarel, I. 2008. Sequential feeding with variations in energy and protein levels improves gait score in meat-type chickens. *Animal* 2(11), 1658-65. doi:10.1017/S1751731108002875.

Lewis, P. D., Danisman, R. and Gous, R. M. 2009. Photoperiodic responses of broilers. III. Tibial breaking strength and ash content. *British Poultry Science* 50(6), 673-9. doi:10.1080/00071660903365612.

Li, H., Wen, X., Alphin, R., Zhu, Z. and Zhou, Z. 2017. Effects of two different broiler flooring systems on production performances, welfare, and environment under commercial production conditions. *Poultry Science* 96(5), 1108-19. doi:10.3382/ps/pew440.

Lien, R. J., Hess, J. B., McKee, S. R., Bilgili, S. F. and Townsend, J. C. 2007. Effect of light intensity and photoperiod on live performance, heterophil-to-lymphocyte ratio, and processing yields of broilers. *Poultry Science* 86(7), 1287-93. doi:10.1093/ps/86.7.1287.

Louton, H., Bergmann, S., Reese, S., Erhard, M., Bachmeier, J., Rösler, B. and Rauch, E. 2018. Animal- and management-based welfare indicators for a conventional broiler strain in 2 barn types (Louisiana barn and closed barn). *Poultry Science* 97(8), 2754-67. doi:10.3382/ps/pey111.

Lund, V. P., Nielsen, L. R., Oliveira, A. R. S. and Christensen, J. P. 2017. Evaluation of the Danish footpad lesion surveillance in conventional and organic broilers: misclassification of scoring. *Poultry Science* 96(7), 2018-28. doi:10.3382/ps/pex024.

Lynch, M., Thorp, B. H. and Whitehead, C. C. 1992. Avian tibial dyschondroplasia as a cause of bone deformity. *Avian Pathology* 21(2), 275-85. doi:10.1080/03079459208418842.

Manangi, M. K., Vazquez-Anon, M., Richards, J. D., Carter, S., Buresh, R. E. and Christensen, K. D. 2012. Impact of feeding lower levels of chelated trace minerals versus industry levels of inorganic trace minerals on broiler performance, yield, footpad health, and litter mineral concentration. *Journal of Applied Poultry Research* 21(4), 881-90. doi:10.3382/japr.2012-00531.

Mandal, R. K., Jiang, T. S., Al-Rubaye, A. A., Rhoads, D. D., Wideman, R. F., Zhao, J., Pevzner, I. and Kwon, Y. M. 2016. An investigation into blood microbiota and its potential association with bacterial chondronecrosis with osteomyelitis (BCO) in broilers. *Scientific Reports* 6, 25882. doi:10.1038/srep25882.

Manning, L., Chadd, S. A. and Baines, R. N. 2007. Water consumption in broiler chicken: a welfare indicator. *World's Poultry Science Journal* 63(1), 63-71. doi:10.1017/S0043933907001274.

Meluzzi, A., Fabbri, C., Folegatti, E. and Sirri, F. 2008a. Survey of chicken rearing conditions in Italy: effects of litter quality and stocking density on productivity, foot dermatitis and carcase injuries. *British Poultry Science* 49(3), 257-64. doi:10.1080/00071660802094156.

Meluzzi, A., Fabbri, C., Folegatti, E. and Sirri, F. 2008b. Effect of less intensive rearing conditions on litter characteristics, growth performance, carcase injuries and meat quality of broilers. *British Poultry Science* 49(5), 509-15. doi:10.1080/00071660802290424.

Muir, W. I. and Groves, P. J. 2018. Incubation and hatch management: consequences for bone mineralization in Cobb 500 meat chickens. *Animal* 12(4), 794-801. doi:10.1017/S1751731117001938.

Müller, B. R., Medeiros, H. A. S., de Sousa, R. S. and Molento, C. F. M. 2015. Chronic welfare restrictions and adrenal gland morphology in broiler chickens. *Poultry Science* 94(4), 574–8. doi:10.3382/ps/pev026.

Nääs, IdA., Paz, I. C. L. A., Baracho, M. S., Menezes, A. G., Bueno, L. G. F., Almeida, I. C. L. and Moura, D. J. 2009. Impact of lameness on broiler well-being. *Journal of Applied Poultry Research* 18(3), 432–9. doi:10.3382/japr.2008-00061.

Nääs, IdA., Paz, I. CdL. A., Baracho, MdS., Menezes, A. Gd, Lima, K. A. Od, Bueno, L. GdF., Mollo Neto, M., Carvalho, V. Cd, Almeida, I. CdL. and Souza, A. Ld 2010. Assessing locomotion deficiency in broiler chicken. *Scientia Agricola* 67(2), 129–35. doi:10.1590/S0103-90162010000200001.

Nagaraj, M., Wilson, C. A. P., Hess, J. B. and Bilgili, S. F. 2007. Effect of high-protein and all-vegetable diets on the incidence and severity of pododermatitis in broiler chickens. *Journal of Applied Poultry Research* 16(3), 304–12. doi:10.1093/japr/16.3.304.

Nicol, C. J. 2015. *The Behavioural Biology of Chickens*. CABI Publishing, Wallingford, UK.

Nicol, C. J., Caplen, G. J., Edgar, J. L. and Browne, W. J. 2009. Associations between welfare indicators and environmental choice in laying hens. *Animal Behaviour* 78(2), 413–24. doi:10.1016/j.anbehav.2009.05.016.

Nielsen, B. L., Thomsen, M. G., Sorensen, J. P. and Young, J. F. 2003. Feed and strain effects on the use of outdoor areas by broilers. *British Poultry Science* 44(2), 161–9. doi:10.1080/0007166031000088389.

Norring, M., Kaukonen, E. and Valros, A. 2016. The use of perches and platforms by broiler chickens. *Applied Animal Behaviour Science* 184, 91–6. doi:10.1016/j.applanim.2016.07.012.

Ohara, A., Oyakawa, C., Yoshihara, Y., Ninomiya, S. and Sato, S. 2015. Effect of environmental enrichment on the behavior and welfare of Japanese broilers at a commercial Farm. *Journal of Poultry Science* 52(4), 323–30. doi:10.2141/jpsa.0150034.

Olukosi, O. A., Olori, V. E., Helmbrecht, A., Lambton, S. and French, N. A. (Eds). 2019. *Poultry Feathers and Skin*. The Poultry Integument in Health and Welfare, CABI Publishing, Wallingford, UK.

Opengart, K., Bilgili, S. F., Warren, G. L., Baker, K. T., Moore, J. D. and Dougherty, S. 2018. Incidence, severity, and relationship of broiler footpad lesions and gait scores of market-age broilers raised under commercial conditions in the southeastern United States. *Journal of Applied Poultry Research* 27(3), 424–32. doi:10.3382/japr/pfy002.

Orth, M. W. and Cook, M. E. 1994. Avian tibial dyschondroplasia – a morphological and biochemical review of the growth-plate lesion and its causes. *Veterinary Pathology* 31(4), 403–4. doi:10.1177/030098589403100401.

Oso, A. O., Idowu, A. A. and Niameh, O. T. 2011. Growth response, nutrient and mineral retention, bone mineralisation and walking ability of broiler chickens fed with dietary inclusion of various unconventional mineral sources. *Journal of Animal Physiology and Animal Nutrition* 95(4), 461–7. doi:10.1111/j.1439-0396.2010.01073.x.

Oviedo-Rondón, E. O. 2019. Holistic view of intestinal health in poultry. *Animal Feed Science and Technology* 250, 1–8. doi:10.1016/j.anifeedsci.2019.01.009.

Oviedo-Rondón, E. O., Wineland, M. J., Funderburk, S., Small, J., Cutchin, H. and Mann, M. 2009a. Incubation conditions affect leg health in large, high-yield broilers. *Journal of Applied Poultry Research* 18(3), 640–6. doi:10.3382/japr.2008-00127.

Oviedo-Rondón, E. O., Wineland, M. J., Small, J., Cutchin, H., McElroy, A., Barri, A. and Martin, S. 2009b. Effect of incubation temperatures and chick transportation conditions on bone development and leg health. *Journal of Applied Poultry Research* 18(4), 671-8. doi:10.3382/japr.2008-00135.

Oznurlu, Y., Sur, E., Ozaydin, T., Celik, I. and Uluisik, D. 2016. Histological and histochemical evaluations on the effects of high incubation temperature on the embryonic development of tibial growth plate in broiler chickens. *Microscopy Research and Technique* 79(2), 106-10. doi:10.1002/jemt.22611.

Pagazaurtundua, A. and Warriss, P. D. 2006. Levels of foot pad dermatitis in broiler chickens reared in 5 different systems. *British Poultry Science* 47(5), 529-32. doi:10.1080/00071660600963024.

Part, C. E., Edwards, P., Hajat, S. and Collins, L. M. 2016. Prevalence rates of health and welfare conditions in broiler chickens change with weather in a temperate climate. *Royal Society Open Science* 3(9), 160197. doi:10.1098/rsos.160197.

Paz, I. C. L. A., Garcia, R. G., Bernardi, R., Seno, LdO., Naas, IdA. and Caldara, F. R. 2013. Locomotor problems in broilers reared on new and re-used litter. *Italian Journal of Animal Science* 12(2). doi:10.4081/ijas.2013.e45.

Petek, M., Sonmez, G., Yildiz, H. and Baspinar, H. 2005. Effects of different management factors on broiler performance and incidence of tibial dyschondroplasia. *British Poultry Science* 46(1), 16-21. doi:10.1080/00071660400023821.

Petek, M., Cibik, R., Yildiz, H., Sonat, F. A., Gezen, S. S., Orman, A. and Aydin, C. 2010. The influence of different lighting programs, stocking densities and litter amounts on the welfare and productivity traits of a commercial broiler line. *Veterinarija ir Zootechnika* 51(73), 36-43.

Puvadolpirod, S. and Thaxton, J. P. 2000. Model of physiological stress in chickens: 4. Digestion and metabolism. *Poultry Science* 79(3), 383-90. doi:10.1093/ps/79.3.383.

Reiter, K. and Bessei, W. 2009. Effect of locomotor activity on leg disorder in fattening chicken. *Berliner Und Munchener Tierarztliche Wochenschrift* 122(7-8), 264-70.

Riddell, C. and Kong, X. M. 1992. The influence of diet on necrotic enteritis in broiler-chickens. *Avian Diseases* 36(3), 499-503. doi:10.2307/1591740.

Roberts, S. J., Cain, R. and Dawkins, M. S. 2012. Prediction of welfare outcomes for broiler chickens using Bayesian regression on continuous optical flow data. *Journal of the Royal Society Interface* 9(77), 3436-43. doi:10.1098/rsif.2012.0594.

Rushton, S. P., Humphrey, T. J., Shirley, M. D. F., Bull, S. and Jorgensen, F. 2009. Campylobacter in housed broiler chickens: a longitudinal study of risk factors. *Epidemiology and Infection* 137(8), 1099-110. doi:10.1017/S095026880800188X.

Rutten, M., Leterrier, C., Constantin, P., Reiter, K. and Bessei, W. 2002. Bone development and activity in chickens in response to reduced weight-load on legs. *Animal Research* 51(4), 327-36. doi:10.1051/animres:2002027.

Saima, M. Z. U., Jabbar, M. A., Ijaz, M. and Qadeer, M. A. 2009. Efficacy of microbial phytase at different levels on growth performance and mineral availability in broiler chicken. *Journal of Animal and Plant Sciences* 19(2), 58-62.

Sandilands, V., Brocklehurst, S., Sparks, N., Baker, L., McGovern, R., Thorp, B. and Pearson, D. 2011. Assessing leg health in chickens using a force plate and gait scoring: how many birds is enough? *Veterinary Record* 168(3), 77. doi:10.1136/vr.c5978.

Sanotra, G. S., Lund, J. D. and Vestergaard, K. S. 2002. Influence of light-dark schedules and stocking density on behaviour, risk of leg problems and occurrence of chronic fear in broilers. *British Poultry Science* 43(3), 344-54. doi:10.1080/0007166012010 36023611.

Sanotra, G. S., Berg, C. and Lund, J. D. 2003. A comparison between leg problems in Danish and Swedish broiler production. *Animal Welfare* 12(4), 677–83.

Sans, E., Federici, J., Dahlke, F. and Molento, C. 2014. Evaluation of Free-Range broilers using the welfare quality® protocol. *Revista Brasileira de Ciência Avícola* 16(3), 297–306. doi:10.1590/1516-635x1603297-306.

Sarica, M., Yamak, U. S. and Boz, M. A. 2014. Effect of production systems on foot pad dermatitis (FPD) levels among slow-, medium- and fast-growing broilers. *European Poultry Science* 78. doi:10.1399/eps.2014.52.

Schmidt, C. J., Persia, M. E., Feierstein, E., Kingham, B. and Saylor, W. W. 2009. Comparison of a modern broiler line and a heritage line unselected since the 1950s. *Poultry Science* 88(12), 2610–9. doi:10.3382/ps.2009-00055.

Scholz-Ahrens, K. E., Ade, P., Marten, B., Weber, P., Timm, W., Asil, Y., Gluer, C. C. and Schrezenmeir, J. 2007. Prebiotics, probiotics, and synbiotics affect mineral absorption, bone mineral content, and bone structure. *Journal of Nutrition* 137(3), 838S–46S. doi:10.1093/jn/137.3.838S.

Schwean-Lardner, K., Fancher, B. I. and Classen, H. L. 2012. Impact of daylength on behavioural output in commercial broilers. *Applied Animal Behaviour Science* 137(1–2), 43–52. doi:10.1016/j.applanim.2012.01.015.

Schwean-Lardner, K., Fancher, B. I., Gomis, S., Van Kessel, A., Dalal, S. and Classen, H. L. 2013. Effect of day length on cause of mortality, leg health, and ocular health in broilers. *Poultry Science* 92(1), 1–11. doi:10.3382/ps.2011-01967.

Sellers, H. S. 2017. Current limitations in control of viral arthritis and tenosynovitis caused by avian reoviruses in commercial poultry. *Veterinary Microbiology* 206, 152–6. doi:10.1016/j.vetmic.2016.12.014.

Senaratna, D., Samarakone, T. S. and Gunawardena, W. W. 2016. Red color light at different intensities affects the performance, behavioral activities and welfare of broilers. *Asian-Australasian Journal of Animal Sciences* 29(7), 1052–9. doi:10.5713/ajas.15.0757.

Shepherd, E. M., Fairchild, B. D. and Ritz, C. W. 2017. Alternative bedding materials and litter depth impact litter moisture and footpad dermatitis. *Journal of Applied Poultry Research* 26(4), 518–28. doi:10.3382/japr/pfx024.

Sherlock, L., McKeegan, D. E. F., Cheng, Z., Wathes, C. M. and Wathes, D. C. 2012. Effects of contact dermatitis on hepatic gene expression in broilers. *British Poultry Science* 53(4), 439–52. doi:10.1080/00071668.2012.707310.

Shim, M. Y., Karnuah, A. B., Anthony, N. B., Pesti, G. M. and Aggrey, S. E. 2012a. The effects of broiler chicken growth rate on valgus, varus, and tibial dyschondroplasia. *Poultry Science* 91(1), 62–5. doi:10.3382/ps.2011-01599.

Shim, M. Y., Karnuah, A. B., Mitchell, A. D., Anthony, N. B., Pesti, G. M. and Aggrey, S. E. 2012b. The effects of growth rate on leg morphology and tibia breaking strength, mineral density, mineral content, and bone ash in broilers. *Poultry Science* 91(8), 1790–5. doi:10.3382/ps.2011-01968.

Siegel, P. B., Gustin, S. J. and Katanbaf, M. N. 2011. Motor ability and self-selection of an analgesic drug by fast-growing chickens. *Journal of Applied Poultry Research* 20(3), 249–52. doi:10.3382/japr.2009-00118.

Simsek, U. G., Dalkilic, B., Ciftci, M., Cerci, I. H. and Bahsi, M. 2009. Effects of enriched housing design on broiler performance, welfare, chicken meat composition and serum cholesterol. *Acta Veterinaria Brno* 78(1), 67–74. doi:10.2754/avb200978010067.

Simsek, U. G., Erisir, M., Ciftci, M. and Tatli Seven, P. 2014. Effects of cage and floor housing systems on fattening performance, oxidative stress and carcass defects in broiler chicken. *Kafkas Universitesi Veteriner Fakultesi Dergisi* 20(5), 727-33.

Skrbic, Z., Pavlovski, Z., Lukic, M. and Petricevic, V. 2015. Incidence of footpad dermatitis and hock burns in broilers as affected by genotype, lighting program and litter type. *Annals of Animal Science* 15(2), 433-45. doi:10.1515/aoas-2015-0005.

Sørensen, P., Su, G. and Kestin, S. C. 2000. Effects of age and stocking density on leg weakness in broiler chickens. *Poultry Science* 79(6), 864-70. doi:10.1093/ps/79.6.864.

Sosnówka-Czajka, E., Skomorucha, I., Herbut, E. and Muchacka, R. 2007. Effect of management system and flock size on the behaviour of broiler chickens. *Annals of Animal Science* 7(2), 329-35.

Stanley, D., Hughes, R. J. and Moore, R. J. 2014. Microbiota of the chicken gastrointestinal tract: influence on health, productivity and disease. *Applied Microbiology and Biotechnology* 98(10), 4301-10. doi:10.1007/s00253-014-5646-2.

Su, G., Sorensen, P. and Kestin, S. C. 1999. Meal feeding is more effective than early feed restriction at reducing the prevalence of leg weakness in broiler chickens. *Poultry Science* 78(7), 949-55. doi:10.1093/ps/78.7.949.

Su, G., Sorensen, P. and Kestin, S. C. 2000. A note on the effects of perches and litter substrate on leg weakness in broiler chickens. *Poultry Science* 79(9), 1259-63. doi:10.1093/ps/79.9.1259.

Sun, Z. W., Yan, L., G, Y. Y., Zhao, J. P., Lin, H. and Guo, Y. M. 2013. Increasing dietary vitamin D-3 improves the walking ability and welfare status of broiler chickens reared at high stocking densities. *Poultry Science* 92(12), 3071-9. doi:10.3382/ps.2013-03278.

Sun, Z. W., Fan, Q. H., Wang, X. X., Guo, Y. M., Wang, H. J. and Dong, X. 2017. High dietary biotin levels affect the footpad and hock health of broiler chickens reared at different stocking densities and litter conditions. *Journal of Animal Physiology and Animal Nutrition* 101(3), 521-30. doi:10.1111/jpn.12465.

Swiatkiewicz, S., Arczewska-Wlosek, A. and Jozefiak, D. 2017. The nutrition of poultry as a factor affecting litter quality and foot pad dermatitis - an updated review. *Journal of Animal Physiology and Animal Nutrition* 101(5), e14-20. doi:10.1111/jpn.12630.

Tahamtani, F. M., Hinrichsen, L. K. and Riber, A. B. 2018. Welfare assessment of conventional and organic broilers in Denmark, with emphasis on leg health. *Veterinary Record* 183(6), 192. doi:10.1136/vr.104817.

Tickle, P. G., Hutchinson, J. R. and Codd, J. R. 2018. Energy allocation and behaviour in the growing broiler chicken. *Scientific Reports* 8(1), 4562. doi:10.1038/s41598-018-22604-2.

Toppel, K., Kaufmann, F., Schon, H., Gauly, M. and Andersson, R. 2019. Effect of pH-lowering litter amendment on animal-based welfare indicators and litter quality in a European commercial broiler husbandry. *Poultry Science* 98(3), 1181-9. doi:10.3382/ps/pey489.

Tullo, E., Fontana, I., Fernandez, A. P., Vranken, E., Norton, T., Berckmans, D. and Guarino, M. 2017. Association between environmental predisposing risk factors and leg disorders in broiler chickens. *Journal of Animal Science* 95(4), 1512-20. doi:10.2527/jas.2016.1257.

Tuyttens, F., Heyndrickx, M., De Boeck, M., Moreels, A., Van Nuffel, A., Van Poucke, E., Van Coillie, E., Van Dongen, S. and Lens, L. 2008. Broiler chicken health, welfare and fluctuating asymmetry in organic versus conventional production systems. *Livestock Science* 113(2-3), 123-32. doi:10.1016/j.livsci.2007.02.019.

van der Pol, C. W., Molenaar, R., Buitink, C. J., van Roovert-Reijrink, I. A. M., Maatjens, C. M., van den Brand, H. and Kemp, B. 2015. Lighting schedule and dimming period in early life: consequences for broiler chicken leg bone development. *Poultry Science* 94(12), 2980-8. doi:10.3382/ps/pev276.

van der Pol, C. W., van Roovert-Reijrink, I. A. M., Aalbers, G., Kemp, B. and van den Brand, H. 2017. Incubation lighting schedules and their interaction with matched or mismatched post hatch lighting schedules: effects on broiler bone development and leg health at slaughter age. *Research in Veterinary Science* 114, 416-22. doi:10.1016/j.rvsc.2017.07.013.

van der Pol, C. W., van Roovert-Reijrink, I. A. M., Maatjens, C. M., Gussekloo, S. W. S., Kranenbarg, S., Wijnen, J., Pieters, R. P. M., Schipper, H., Kemp, B. and van den Brand, H. 2019. Light-dark rhythms during incubation of broiler chicken embryos and their effects on embryonic and post hatch leg bone development. *PLoS ONE* 14(1), e0210886. doi:10.1371/journal.pone.0210886.

Van Hertem, T., Norton, T., Berckmans, D. and Vranken, E. 2018. Predicting broiler gait scores from activity monitoring and flock data. *Biosystems Engineering* 173, 93-102. doi:10.1016/j.biosystemseng.2018.07.002.

Vargas-Galicia, A. J., Sosa-Montes, E., Rodriguez-Ortega, L. T., Pro-Martinez, A., Ruiz-Feria, C. A., González-Cerón, F., Gallegos-Sánchez, J., Arreola-Enríquez, J. and Bautista-Ortegac, J. 2017. Effect of litter material and stocking density on bone and tendon strength, and productive performance in broilers. *Canadian Journal of Animal Science* 97(4), 673-82.

Vasdal, G., Vas, J., Newberry, R. C. and Moe, R. O. 2019. Effects of environmental enrichment on activity and lameness in commercial broiler production. *Journal of Applied Animal Welfare Science* 22(2), 197-205. doi:10.1080/10888705.2018.1456339.

Ventura, B. A., Siewerdt, F. and Estevez, I. 2010. Effects of barrier perches and density on broiler leg health, fear, and performance. *Poultry Science* 89(8), 1574-83. doi:10.3382/ps.2009-00576.

Ventura, B. A., Siewerdt, F. and Estevez, I. 2012. Access to barrier perches improves behavior repertoire in broilers. *PLoS ONE* 7(1), e29826. doi:10.1371/journal.pone.0029826.

Villagrá, A., Olivas, I., Benitez, V. and Lainez, M. 2011. Evaluation of sludge from paper recycling as bedding material for broilers. *Poultry Science* 90(5), 953-7. doi:10.3382/ps.2010-00935.

Villarroel, M., Francisco, I., Ibanez, M. A., Novoa, M., Martinez-Guijarro, P., Méndez, J. and De Blas, C. 2018. Rearing, bird type and pre-slaughter transport conditions of broilers II. Effect on foot-pad dermatitis and carcass quality. *Spanish Journal of Agricultural Research* 16(2), e0504. Available at: http://revistas.inia.es/index.php/sjar/article/view/12015/4038 (accessed on 19 June 2019). doi:10.5424/sjar/2018162-12015.

Waldenstedt, L. 2006. Nutritional factors of importance for optimal leg health in broilers: a review. *Animal Feed Science and Technology* 126(3-4), 291-307. doi:10.1016/j.anifeedsci.2005.08.008.

Weeks, C. A. and Butterworth, A. 2004. *Measuring and Auditing Broiler Welfare* (vol. 1). CABI Publishing, Wallingford, UK.

Weeks, C. A., Danbury, T. D., Davies, H. C., Hunt, P. and Kestin, S. C. 2000. The behaviour of broiler chickens and its modification by lameness. *Applied Animal Behaviour Science* 67(1–2), 111–25. doi:10.1016/s0168-1591(99)00102-1.

Weeks, C. A., Knowles, T. G., Gordon, R. G., Kerr, A. E., Peyton, S. T. and Tilbrook, N. T. 2002. New method for objectively assessing lameness in broiler chickens. *Veterinary Record* 151(25), 762–4.

Weimer, S. L., Wideman, R. F., Scanes, C. G., Mauromoustakos, A., Christensen, K. D. and Vizzier-Thaxton, Y. 2019. The utility of infrared thermography for evaluating lameness attributable to bacterial chondronecrosis with osteomyelitis. *Poultry Science* 98(4), 1575–88. doi:10.3382/ps/pey538.

Welfare Quality. 2009. Welfare Quality assessment protocol for poultry (broilers, laying hens). Welfare Quality Consortium, Lelystad, the Netherlands. Available at: http://www.welfarequalitynetwork.net/downloadattachment/45627/21652/Poultry%20Protocol.pdf (accessed on 19 June 2019).

Whitehead, C. C., McCormack, H. A., McTeir, L. and Fleming, R. H. 2004. High vitamin D-3 requirements in broilers for bone quality and prevention of tibial dyschondroplasia and interactions with dietary calcium, available phosphorus and vitamin A. *British Poultry Science* 45(3), 425–36. doi:10.1080/00071660410001730941.

Wideman, R. F. 2016. Bacterial chondronecrosis with osteomyelitis and lameness in broilers: a review. *Poultry Science* 95(2), 325–44. doi:10.3382/ps/pev320.

Wideman, R. F. and Prisby, R. D. 2013. Bone circulatory disturbances in the development of spontaneous bacterial chondronecrosis with osteomyelitis: a translational model for the pathogenesis of femoral head necrosis. *Frontiers in Endocrinology* 3, 183. doi:10.3389/fendo.2012.00183.

Wideman, R. F., Hamal, K. R., Stark, J. M., Blankenship, J., Lester, H., Mitchell, K. N., Lorenzoni, G. and Pevzner, I. 2012. A wire-flooring model for inducing lameness in broilers: evaluation of probiotics as a prophylactic treatment. *Poultry Science* 91(4), 870–83. doi:10.3382/ps.2011-01907.

Wideman, R. F., Al-Rubaye, A., Reynolds, D., Yoho, D., Lester, H., Spencer, C., Hughes, J. D. and Pevzner, I. Y. 2014. Bacterial chondronecrosis with osteomyelitis in broilers: influence of sires and straight-run versus sex-separate rearing. *Poultry Science* 93(7), 1675–87. doi:10.3382/ps.2014-03912.

Wideman, R. F., Blankenship, J., Pevzner, I. Y. and Turner, B. J. 2015. Efficacy of 25-OH vitamin D-3 prophylactic administration for reducing lameness in broilers grown on wire flooring. *Poultry Science* 94(8), 1821–7. doi:10.3382/ps/pev160.

Wijesurendra, D. S., Chamings, A. N., Bushell, R. N., Rourke, D. O., Stevenson, M., Marenda, M. S., Noormohammadi, A. H. and Stent, A. 2017. Pathological and microbiological investigations into cases of bacterial chondronecrosis and osteomyelitis in broiler poultry. *Avian Pathology* 46(6), 683–94. doi:10.1080/03079457.2017.1349872.

Wijtten, P. J. A., Hangoor, E., Sparla, J. K. and Verstegen, M. W. A. 2010. Dietary amino acid levels and feed restriction affect small intestinal development, mortality, and weight gain of male broilers. *Poultry Science* 89(7), 1424–39. doi:10.3382/ps.2009-00626.

Xavier, D. B., Broom, D. M., McManus, C. M. P., Torres, C. and Bernal, F. E. M. 2010. Number of flocks on the same litter and carcase condemnations due to cellulitis, arthritis and contact foot-pad dermatitis in broilers. *British Poultry Science* 51(5), 586–91. doi:10.1080/00071668.2010.508487.

Yalçin, S., Molayoglu, H. B., Baka, M., Genin, O. and Pines, M. 2007. Effect of temperature during the incubation period on tibial growth plate chondrocyte differentiation and the incidence of tibial dyschondroplasia. *Poultry Science* 86(8), 1772–83. doi:10.1093/ps/86.8.1772.

Yamak, U. S., Sarica, M., Boz, M. A. and Ucar, A. 2016. Effect of reusing litter on broiler performance, foot-pad dermatitis and litter quality in chickens with different growth rates. *Kafkas Universitesi Veteriner Fakultesi Dergisi* 22(1), 85–91.

Yan, F. F., Wang, W. C. and Cheng, H. W. 2018. Bacillus subtilis based probiotic improved bone mass and altered brain serotoninergic and dopaminergic systems in broiler chickens. *Journal of Functional Foods* 49, 501–9. doi:10.1016/j.jff.2018.09.017.

Yan, F. F., Mohammed, A. A., Murugesan, G. R. and Cheng, H. W. 2019. Effects of a dietary synbiotic inclusion on bone health in broilers subjected to cyclic heat stress episodes. *Poultry Science* 98(3), 1083–9. doi:10.3382/ps/pey508.

Yang, H. M., Xing, H., Wang, Z. Y., Xia, J. L., Wan, Y., Hou, B. and Zhang, J. 2015. Effects of intermittent lighting on broiler growth performance, slaughter performance, serum biochemical parameters and tibia parameters. *Italian Journal of Animal Science* 14(4), 4143. doi:10.4081/ijas.2015.4143.

Yildiz, H., Gunes, N., Cengiz, S., Ozcan, R., Petek, M., Yilmaz, B. and Arican, I. 2009. Effects of ascorbic acid and lighting schedule on tibiotarsus strength and bone characteristics in broilers. *Archiv fur Tierzucht* 52(4), 432–44.

Yildiz, A., Yildiz, K. and Apaydin, B. 2014. The effect of vermiculite as litter material on some health and stress parameters in broilers. *Kafkas Universitesi Veteriner Fakultesi Dergisi* 20(1), 129–34.

Zhao, J., Shirley, R. B., Vazquez-Anon, M., Dibner, J. J., Richards, J. D., Fisher, P., Hampton, T., Christensen, K. D., Allard, J. P. and Giesen, A. F. 2010. Effects of chelated trace minerals on growth performance, breast meat yield, and footpad health in commercial meat broilers. *Journal of Applied Poultry Research* 19(4), 365–72. doi:10.3382/japr.2009-00020.

Zhao, Z. G., Li, J. H., Li, X. and Bao, J. 2014. Effects of housing systems on behaviour, performance and welfare of fast-growing broilers. *Asian-Australasian Journal of Animal Sciences* 27(1), 140–6. doi:10.5713/ajas.2013.13167.

Zikic, D., Djukic-Stojcic, M., Bjedov, S., Peric, L., Stojanovic, S. and Uscebrka, G. 2017. Effect of litter on development and severity of foot pad dermatitis and behavior of broiler chickens. *Revista Brasileira de Ciência Avícola* 19(2), 247–54. doi:10.1590/1806-9061-2016-0396.

Chapter 3

Broiler breeding flocks: management and animal welfare

Ingrid C. de Jong and Rick A. van Emous, Wageningen Livestock Research, The Netherlands

1 Introduction

This chapter discusses the management of broiler breeders, that is, the parent stock of broiler chickens, and welfare issues related to the different life stages of broiler breeders. Where commercial broilers are the product of a cross of four lines, broiler parent stock (broiler breeders) are the product of a cross of two lines; both the male and female broiler breeder is the product of a cross of a specific paternal line and a specific maternal line (Hiemstra and Ten Napel, 2013). The actual figures on the number of broiler breeders worldwide are lacking; as an indication, in Europe the number of broiler breeders is estimated to be 44 million (Horne and Bondt, 2014). Management and housing of grandparents and great grandparents (but not pedigree stock) is to a large extent similar to that of broiler breeders (EFSA, 2010; Hiemstra and Ten Napel, 2013) and not described in this chapter. Nowadays, three companies dominate the world market for broiler breeding stock: Aviagen Broiler Breeders, Cobb-Vantress and Hubbard.

http://dx.doi.org/10.19103/AS.2016.0011.26

Published by Burleigh Dodds Science Publishing Limited, 2017.

The majority of the broiler breeders worldwide are the parent stock of the so-called standard or fast growing broilers, which reach a body weight of 2.5 kg in 42 days or less (EFSA, 2010). Dwarf parental females are used to produce broilers of intermediate (2.2 kg in 56-63 days of age) or slow growth rate (2.2 kg in 70-80 days of age) (De Jong and Swalander, 2013). Although worldwide only a small percentage of parent stock for intermediate or slow growing broilers is housed, in some countries a larger proportion of the total number of broiler breeders produce intermediate or slower growing broiler strains. For example, in France, the majority of the parent stock –85% according to De Jong and Guemene (2011) – are parents of intermediate or slower growing broiler strains (EFSA, 2010). In Europe, about 8% of the broiler breeders are estimated to be parent stock of intermediate or slower growing broilers (Horne and Bondt, 2014).

We provide a short, general description of housing and management of broiler breeders during both the rearing and the production period in the next paragraph. In addition, we focus on (major) welfare issues related to the management of broiler breeders and the current state-of-the-art research related to these welfare issues.

2 Housing conditions and management in the rearing period

In the rearing period, which comprises the age between 0 and 18-22 weeks, broiler breeder pullets are prepared for the production phase. The aim of the rearing period is to produce birds of ideal weight, uniformity, condition and stage of sexual maturity when they enter the production house (Leeson and Summers, 2000). Body weight and flock uniformity are important production indicators during the rearing period (Zuidhof et al., 2015).

Although there is worldwide variation in housing conditions during the rearing period due to differences in legislation, climatic conditions and labour costs, layout of rearing houses does not vary to a large extent. During the rearing period, broiler breeders are usually housed on a fully littered floor, and litter types vary between regions due to availability. In Europe, wood shavings, peat and straw are commonly used. Cage rearing is not common (De Jong and Swalander, 2013; EFSA, 2010). Feed is either distributed via pan feeders, chain feeders or spin feeders (systems that spread the feed in the litter) and can be provided as mash, crumb or pellet. Pelleted feed should be used in combination with spin feeders. Parent stock of fast growing broiler chickens is subjected to feed restriction during the rearing period (De Jong and Guemene, 2011; EFSA, 2010). In addition, water provision may be restricted (Hocking et al., 1993). Feed can be either provided daily or a skip-a-day feeding regime may be applied (EFSA, 2010). Skip-a-day feeding regimes can be in the form of 6/1, 5/2 or 4/3 feeding programmes (1, 2 or 3 days without feed

Published by Burleigh Dodds Science Publishing Limited, 2017.

Figure 1 Example of a broiler breeder rearing house in northwest Europe (photo provided by N. Katier).

each week and a larger portion on the feeding days). In Europe, usually daily feeding applied as legislation does not allow skip-a-day feeding programmes. Either raised platforms or perches are provided, usually from about 6 weeks of age onwards, to train the birds to use the higher levels in a house during the production period.

Depending on the region, climate and legislation, houses can be fully climate controlled and without daylight entrance or with windows, or either open-sided with mesh curtains. Males and females are reared separately so that different feeding programmes can be applied in rearing. Stocking density also varies between countries, depending on the climatic conditions and also on country-specific legislation. In European countries, stocking densities between 7 and 10 birds/m^2 (females) or 4 and 8 birds/m^2 (males) are generally applied, with lower stocking densities in open-sided houses (EFSA, 2010). After a few days, usually a light programme with 8 h light per day is applied with a light intensity between 10 and 20 lux, although lower light intensities may be applied in case of injurious pecking or to reduce birds' activity (EFSA, 2010). The management guides of the breeding companies are usually taken as the basis for stocking density, light programme, feeding programme and so on (e.g. Aviagen, 2013; Cobb, n.d.; Hubbard, n.d.). Figure 1 shows an example of a rearing house in northwest Europe.

3 Housing conditions and management in the production period

The main goal in broiler breeder production is to provide fertilised eggs to produce a maximum number of healthy and robust day-old broiler chicks

(Zuidhof et al., 2007). Important in relation to the management of adult breeders is maintaining the health status of the flock while keeping the egg production at a high level. Major criteria for monitoring birds for management purposes include body weight, body condition, egg production and hatching, hatchability and infertility and egg weight (Leeson and Summers, 2000).

The transition from the rearing period to the production period involves transportation to the production house, usually at a separate production farm. The production period starts between 18 and 22 weeks and lasts until 60-65 weeks, depending on the performance of a breeder flock. Males and females are reared separately and mixed at the beginning of the production period, and in Europe usually arrive on the same day at the production farm (EFSA, 2010; Van Tuijl, Aviagen, 2016, pers. comm.). The majority of the breeders are housed in floor house systems during the production period. Houses may be artificially lit, with or without windows or open-sided with curtains, which is dependent on legislation, region and climate. Artificial light in addition to natural light can be used to stimulate reproduction. Although less common, broiler breeders can also be housed in multi-tier colony cages during the production period. These are furnished cages with laying nests, perches and also a small litter area (De Jong and Swalander, 2013). Aviaries are not used for broiler breeders. During the production period broiler breeders can also be housed in conventional cages where artificial insemination is applied; this is uncommon in Europe, but may be found in other parts of the world (EFSA, 2010).

Floor house systems consist of a litter area and a certain proportion of slatted floors from which the nests can be accessed. The proportion of littered floor versus raised slatted floor (plastic or wooden slats) may differ between countries and regions, and the layout is also regionally dependent. Water is usually supplied on the slatted area and feed may be provided on the litter and/or on the slatted area. Litter types may vary between countries, for example, wood shavings, straw or peat. Feed can be provided in various forms (mash, pellet and crumbs) via feeder tracks or pans containing a male exclusion system. Male breeders are usually fed via trough feeding or feeder pans near the walls of the house, at such a height that the female breeders cannot reach there. Water can be provided via nipples, bell drinkers or cups. Feed restriction is applied during the production period to control the body weight, but the restriction level is much less severe as compared to the rearing period (De Jong and Jones, 2006). Water restriction may also be applied (EFSA, 2010). Figure 2 shows an example of a production house in northwest Europe.

Stocking density during the production period varies between 5 and 7.5 birds/m² depending on the legislation and region. Stocking density is usually lower in open-sided houses (EFSA, 2010). The percentage of males at the start of the production period is between 8 and 11%, and this decreases due to selection and mortality of males. At the start of the production around 23

Figure 2 Example of a broiler breeder production house. Laying nests and slatted floors are present in the middle of the house (left side of the picture) and the litter area is at both sides of the slatted area near the walls. This house does not have windows (Photo: R. van Emous).

weeks of age, 7.5-9% males are present in a flock (EFSA, 2010). The selection criterion of male breeders includes absence of mating activity and health problems (e.g. leg problems). About 15-25% of the males are selected during the production period. In some countries, 'spiking' of males is common practice. Inactive males are removed from the flock and replaced by younger and more active males to maintain the production of fertile eggs at a high level (Leeson and Summers, 2000). However, spiking involves the risk of introduction of pathogens and it may be stressful to the birds because male aggression may increase (EFSA, 2010). Sometimes intra-spiking (swapping older males at the same farm from one house to the next) is used as a method to increase male activity and thus fertility, which has a much lower biosecurity risk (Casanovas, 2000).

4   Welfare issues: restricted feeding and water restriction

4.1 Restricted feeding

Parent stock of the so-called fast growing broilers (broilers that will reach a body weight of 2.5 kg in about 6 weeks) is subjected to restricted feeding programmes so that the birds will grow according to the growth curve as indicated by the breeding company. If broiler breeders are fed unrestricted, they will grow very fast, reach high body weights and as a result develop health (e.g. leg problems and increased mortality) (Heck et al., 2004; Mench, 2002) and reproduction problems. Low egg production associated with multiple ovulations, accelerated sexual maturity, low egg quality and low persistency of lay have been reported in unrestricted fed broiler breeders (Heck et al., 2004; Hocking et al., 1989), which is unfavourable with respect to the sustainability of the breeding flock. The high growth rate when fed unrestricted is due to the genetic selection for fast and efficient growth of standard, fast growing broiler chickens (Dawkins and Layton, 2012; Renema et al., 2007). Parent stock of so-called slower growing broiler strains are usually subjected to different, less restricted or unrestricted feeding programmes, as these birds have (much) slower growth rates due to a different genetic background (De Jong and Guemene, 2011). Parent stock of these so-called slower or intermediate growing broilers may either be a combination of a slower growing, usually dwarf female and a standard, fast growing male (that is subjected to a restricted feeding programme) or a combination of both a slower growing male and female.

Researchers in the early 1990s described the effect of the restricted feeding programme in broiler breeders on the welfare of the birds. Restricted fed broiler breeders show behavioural signs of stress and frustration, that is, redirected oral behaviours resulting in stereotypic object pecking (and if directed at the drinker resulting in overdrinking), hyperactivity, pacing (stereotypic walking) (De Jong et al., 2002; Hocking, 1993; Hocking, Maxwell and Mitchell, 1993; Hocking et al., 1996; Savory and Maros, 1993; Savory et al., 1992, 1993, 1996), increased aggression (Jones et al., 2004) and increased feeding motivation (Savory and Lariviere, 2000). It has also been reported that physiological indicators of (chronic) stress were observed, such as increased plasma corticosterone levels and increased heterophil:lymphocyte ratios in the blood (De Jong et al., 2003; Hocking, 1993; Hocking, Maxwell and Mitchell, 1996; Savory and Mann, 1997). However, more recently it has been debated if these physiological indicators indeed reflect stress, frustration and/or hunger in the birds or merely reflect the metabolic state (D'Eath et al., 2009; De Jong et al., 2003). Research continued with the aim to find reliable methods to measure hunger and motivation for appetitive behaviour in broiler breeders. A test was developed to measure appetitive behaviour (Dixon et al., 2014), but thus far no reliable behavioural test to assess hunger has been developed (Buckley et al., 2011a,b, 2015). Nevertheless, there is little doubt in research that animals undergoing quantitative feed restriction are hungry and that this negatively affects the welfare of the bird (D'Eath et al., 2009).

During the first 2-3 weeks of the rearing period, feed is provided unrestricted and thereafter a restricted feeding programme is started. The period in which the most severe restriction is applied starts around 6-7 weeks of age and runs to 15-16 weeks of age (De Jong and Jones, 2006). Restriction levels were estimated to be 25-33% of the intake of broiler breeders fed *ad libitum* (De Jong et al., 2002), but there are no recent data on relative restriction level. It can be expected that due to the continuing selection for efficient growth of the progeny, the relative restriction level has been increased since then (Zuidhof et al., 2014). By the end of the rearing period, that is, from 16 weeks of age onwards, the daily feed allowance slightly increases to prepare the birds for the production period. In addition, during the production period restricted feeding is applied, but restriction levels are much less severe as compared to the rearing period. Restriction levels of 45-80% of the *ad libitum* intake are applied until the peak of lay (Bruggeman et al., 1999) and restriction levels to about 80% of *ad libitum* intake are applied after peak of lay (Hocking et al., 2002), although also here no recent data are available.

The majority of research with respect to welfare of broiler breeders is related to the effects of restricted feeding on welfare. More recently, research focused on various methods to reduce the negative effects of restricted feeding such as using alternative feed composition or alternative feeding methods. The results of these studies are discussed below.

4.1.1 Fibre diluted diets or low protein diets in the rearing and the production period

It has been shown that increasing the fibre contents of the diet or decreasing the energy and protein contents of the diet can have positive effects on the welfare of broiler breeders, as measured by a reduction in time spent on behaviours indicative of frustration, stress or hunger. Hocking et al. (2004) applied diets diluted with 50, 100 and 200 g/kg ground oat hulls, ground unmolassed sugar beet pulp or sunflower meal and showed that diets with the highest concentration of oat hulls and sugar beet pulp had positive effects on behaviour during the rearing period (decreased prevalence of damaging pecking behaviour). The diets diluted with sugar beet pulp were associated with higher water content in the gastrointestinal tract and the authors suggested that these diets improved satiety and thus had the largest effect on the welfare of broiler breeders (Hocking et al., 2004).

De Jong et al. (2005a) applied diets with increased fibre contents and various fibre types in the rearing and the production period (8.4 MJ/kg (standard diet) vs 9.2 and 10.4 MJ/kg diets in the rearing period, and 11.7 MJ/kg (standard diet) vs 9.2 and 10.5 MJ/kg during the production period). They observed that the diet of 8.4 MJ/kg had some positive effects on the behaviour during the first half of the rearing period (decreased prevalence of stereotypic object pecking), but not during the second half of the rearing period, and even seemed to increase stress during the production period (indicated by a higher heterophil:lymphocyte ratio in the blood in the production period). Sandilands et al. (2005) and Tolkamp et al. (2005) fed broiler breeders *ad libitum* with diets diluted with oat hulls but in addition used an appetite suppressant (400 g/kg oat hulls or 400 g/kg oat hulls and 24–110 g/kg calcium propionate added to the standard diet, depending on bird's age) or used calcium propionate in combination with low protein content and provided these diets *ad libitum* (150 g/kg crude protein vs 200 g/kg crude protein in the standard diet vs diets diluted with oat hulls or oat hulls in combination with calcium propionate (Sandilands et al., 2006)). Stereotypic pecking was virtually absent, time spent sitting significantly increased and feeding motivation was reduced when applying diets diluted with oat hulls and including an appetite suppressant, indicating that welfare of broiler breeders during the rearing period might be improved (Sandilands et al., 2005; Tolkamp et al., 2005). The same effects on behaviour during the rearing period were found when appetite suppressants were used in combination with low protein diets (Sandilands et al., 2006). Morrissey et al. (2014b) showed that diets with increased fibre content (soy bean hulls, 400 g/kg) and an appetite suppressant (calcium propionate, 10–50 g/kg) had a positive effect on the behaviour of broiler breeders in the rearing period. However, it can be questioned if the use of appetite suppressants will be accepted by society; in addition, researchers questioned if the effects of appetite suppressants are caused by the fact that birds feel ill (Hocking and Bernard, 1993), which is unfavourable in terms of animal welfare.

Nielsen et al. (2011) tested the effects of diets with a high proportion of soluble or insoluble fibre. Three diets were applied, a commercial diet with non-starch polysaccharides (NSP) as fibre content (80% insoluble fibres) and two diets with twice the fibre content and a higher (89%) or lower (71%) proportion of insoluble fibres. They found that the experimental diet with twice the fibre amount and the high proportion of insoluble fibres had a positive effect on the behaviour of broiler breeders in the rearing period; stereotypic pecking was absent and more dust bathing, comfort behaviour and foraging was observed in these birds as compared to the control diet and the diet with soluble fibres. The diet with the low proportion of insoluble fibres negatively affected the litter quality (increased moisture level as compared to the other treatment groups) and birds fed these diets showed behavioural signs of discomfort. Thus, this diet was not preferred in terms of broiler breeder welfare (Nielsen et al., 2011). Others tested diets diluted with insoluble and soluble fibres (cellulose, wheat bran and cottonseed meal) during the production period and found that not only eating time increased, but that for the diets with insoluble fibres (cellulose and wheat bran) egg production also increased and plasma corticosterone concentrations decreased as compared to the diet diluted with cottonseed meal (Moradi et al., 2013), indicating a positive effect of these types of diluted diets on broiler breeder welfare and production.

Van Emous et al. (2014, 2015a) investigated the effects of three dietary protein levels and different growth patterns (2400 g vs 2200 g at 20 weeks of age) on indicators of welfare during the rearing and the production period.

Published by Burleigh Dodds Science Publishing Limited, 2017.

A positive effect on welfare indicators was found for the low dietary protein level diet in the rearing period, but not of the high growth pattern; stereotypic object pecking in the rearing period was significantly reduced in the low dietary protein treatment as compared to the other diets (van Emous et al., 2014, 2015a). When these diets in the rearing period were followed by a low, standard or high energy diet in the production period, it was shown that the low energy diet had a slightly positive effect on behavioural indicators of stress during the production period as compared to the medium or high energy diet, although both the low energy and low protein diet had negative effects on the quality of the feather cover. The authors concluded that increasing the energy:protein ratio had a positive effect on broiler breeder welfare as indicated by the behaviour of the pullets, especially in the rearing period (van Emous, 2015a).

To conclude, modifications of the diet composition by increasing the fibre level and/or increasing the energy:protein ratio are promising to alleviate the effects of feed restriction on broiler breeder welfare, especially during the rearing period. This research area needs further attention to promote application of these types of feed in practice.

4.1.2 Feeding programmes, feeding frequencies and feeding methods

In Europe daily feeding is commonly applied, whereas in North America skip-a-day feeding programmes are applied as these improve flock uniformity (de Beer and Coon, 2007). A few studies focused on the effects of daily or skip-a-day feeding programmes during the rearing period on broiler breeder welfare. Morrissey et al. (2014b) did not find evidence that either skip-a-day feeding (feeding on alternate days) or daily feeding differed in the extent to which broiler breeders were hungry or frustrated during the rearing period, although during the production period there seemed to be a positive effect of skip-a-day feeding over daily feeding (improved feather condition ascribed to less feather pecking) (Morrissey et al., 2014a). Skinner-Noble and Teeter (2009, cited in (EFSA, 2010)) reported no differences in stress levels in birds fed either daily or as per the skip-a-day feeding programme.

De Jong et al. (2005b) tested whether increasing the feeding frequency from once to twice a day, or scattering feed in the litter would reduce stress or hunger in the rearing period. They did not find any positive effects of both these methods on the behaviour and physiological indicators of stress. In general, it is advised to use mash feed instead of pellets to increase feeding time in broiler breeders, although this has not any effect on feelings of hunger. Furthermore, it is important to provide sufficient space at the feeders to prevent aggression in the birds at the time of feeding, as aggression around feeding may result in increased levels of feather and skin damage. Finally, the speed to which feed is distributed is very important to promote equal feed consumption between individuals, to promote flock uniformity and to prevent aggression between birds.

4.1.3 Feeding of males

Studies on the effect of feed restriction on broiler breeder welfare usually focus on the females, as there are numerically much more females present than male broiler breeders. According to Renema et al. (2007), males are subjected to a less severely restricted feeding regime during the rearing period as compared to females, but males are relatively more restricted during the production period as compared to females (EFSA, 2010). It is generally considered that feed restriction affects male welfare to a similar extent as the females; however, research is needed to confirm this hypothesis. De Jong et al. (2011) observed male and female broiler breeder behaviour during the rearing period when housed at two stocking densities (standard vs reduced stocking density), and observed that general behaviour of the males differed from female behaviour at both stocking densities. Object pecking was less common in male breeders as compared to female breeders during the rearing period, and males showed more standing and walking behaviour as compared to females. It is not clear if these differences in general behaviour imply a different stress level in males and females or just express differences in general behaviour between both sexes.

4.1.4 Parent stock of slower growing broilers

Substitution of the standard broiler breeders with dwarf or slower growing broiler breeders may be a strategy to improve welfare of broiler breeders during the rearing period (Decuypere et al., 2006). Jones et al. (2004) showed that replacing standard broiler breeders with dwarf female breeders favourably changed the behaviour (e.g. less stereotypic pecking), because the degree of food restriction during the rearing period was much less in the dwarf breeders as compared to the standard breeders. The dwarf breeders concern only one sex (only females), but numerically many more broiler breeder females are subjected to feed restriction as compared to males, and thus using dwarf females may be a solution for a large part of the broiler breeder population to improve welfare. However, using slower growing or dwarf females corresponds only to a specific market demand and in practice these bird types are not used on a large scale (De Jong and Guemene, 2011).

4.2 Water restriction

Farmers may apply water restriction during the rearing and the production period (starting from a few weeks of age onwards), although in some countries this is forbidden (EFSA, 2010). Water restriction is usually applied to prevent overdrinking in restricted fed broiler breeders, which may result in wet litter and, in severe cases, in polydipsia (Hocking et al., 1993). When water restriction is applied, water is usually provided around feeding and during a few hours after feeding, and possibly on other occasions during the day (EFSA, 2010). Water restriction is generally considered unfavourable with respect to welfare, although Hocking et al. (1993) showed that limited access to water during the rearing period of broiler breeders did not affect welfare indicators.

5   Welfare issues: excisions, mating behaviour and quality of feather cover

5.1 Excisions (beak trimming, toe clipping and de-spurring)

Depending on country-specific legislation, excisions can be carried out in broiler breeders. These are beak trimming (which can be applied in both males and females), toe clipping and de-spurring. EFSA (2010) reported that comb dubbing is only practised in less than 10% of the breeder population and not recommended by the breeders. Excisions can be carried out at the hatchery (de-spurring, toe clipping or beak trimming (applying hot blade) or infrared beak treatment (Gentle and McKeegan, 2007) or on-farm (hot blade beak trimming). Although toe clipping is commonly applied (usually the backward or inner toe is trimmed), de-spurring is not applied in all breeds (EFSA, 2010) and beak trimming is prohibited in some EU countries (Fiks-van Niekerk et al., 2009).

Both de-spurring and toe clipping are carried out to prevent feather and skin damage in the females due to mating behaviour of the males. Sharp and long spurs and toes may damage the feathers and subsequently the skin, and in flocks with non-mutilated males this can result in severe wounds in the females (and as a consequence increased mortality and a negative effect on egg production and fertility). This has especially been reported for males with intact toes (De Jong, 2016, pers. comm.). Beak trimming is performed in female and male broiler breeders to prevent feather and skin damage due to injurious pecking behaviour. Beak trimming in males also reduces feather damage at the back of the head, where the male grabs the female with his beak while mating (Gentle and McKeegan, 2007). However, beak trimming in males and females is not practised in all countries and in general injurious pecking (feather pecking and resulting cannibalism) is not very common in broiler breeder flocks with intact beaks. Specific management might be required, such as providing pecking substrates (bales, pecking blocks) or, in case of an outbreak of injurious pecking, managing the light intensity.

In a pilot study on a commercial broiler breeder farm, it was found that females (Ross 308) with intact beaks had better performance during the rearing period than infrared beak-trimmed females. This was due to the fact that there was no negative effect of beak treatment on feed and water intake during the first days on the rearing farm, resulting in improved growth and uniformity and less mortality in the non-beak-trimmed as compared to the beak-trimmed flocks. Feather scoring during the rearing and the production period showed that in general there was no negative effect of non-beak trimming on the feather cover in flocks with intact beaks (De Jong et al., 2013). A pilot study in commercial flocks with non-beak-trimmed and beak-trimmed males (Ross 308; both housed with non-beak-trimmed Ross 308 females) showed that non-beak trimming of male broiler breeders also did not result in any negative effects on feather damage in females and males, nor was there an effect on the technical results of these flocks (De Jong, Gunnink and Van Emous, 2016, pers. comm.). It is currently unknown if there are differences between broiler breeder lines in the propensity to develop injurious pecking behaviour.

5.2 Mating behaviour

Mating behaviour of both males and females is important with respect to maximising the production of fertile eggs. However, broiler breeder males may show rough, aggressive behaviour towards females during mating. This rough male behaviour may lead to feather and skin damage and fearfulness in females. As a result, females may hide in the nests and fertility will be negatively affected (Millman et al., 2000). One of the reasons for male aggression may be that males reach maturity earlier than females, leading to forced copulations and distress in the females, which in turn results in females hiding on the slats and in the nests (Leone et al., 2007). However, also with proper management in relation to sexual development of males and females, rough and aggressive mating behaviour of the males has been observed. In addition, it was observed that courtship behaviour in males was virtually absent (De Jong et al., 2009; Hocking and Bernard, 2000; Jones and Prescott, 2000; Millman et al., 2000). It has also been observed that females do not seem to respond properly (with crouching behaviour) to male approach (De Jong et al., 2009).

It is unknown why male broiler breeders show this rough behaviour towards females during mating. Rough mating behaviour was not related to aggressive behaviour per se (Millman and Duncan, 2000b) and feed restriction in the rearing period was not related to rough mating behaviour (Millman and Duncan, 2000a). Genetic background may play a role, as males from laying strains responded less aggressive during mating than broiler breeder males (Millman and Duncan, 2000b). Reducing the stocking density, thus, providing males with more space to perform courtship behaviour, indeed had a positive effect on the quality of the mating behaviour (more courtship behaviour and more successful matings) and also resulted in a higher egg production, more fertile eggs and a higher number of chicks per hen (De Jong et al., 2011). Jones et al. (2001) showed that enriching the light conditions with UV_A improved the transmission of sexual signals and thus the quality of the mating behaviour.

It has been estimated that mating frequency is 5-10 times higher in a broiler breeder house as compared to flocks of chickens housed under natural conditions (Van Emous, 2010). This relatively high frequency of mating may affect the relationship between the males and females, resulting in females avoiding males as hypothesised by Fontana et al. (1992). Over-mating might be avoided by separating females and males temporarily during the day. Based on this hypothesis, a new housing system for broiler breeders, called the Quality Time® Concept (QTC), has been developed (Fig. 3) (Van Emous, 2010). Males are separated from females during 5 hours/day, using a separate feeding system and a moving fence. After a successful pilot experiment, two on-farm experiments were carried out in a broiler breeder house with 15 000 birds. The house was divided in six compartments. In the QTC compartments more voluntary and successful matings were observed. In addition, quality of the sexual behaviour improved, which resulted in an improved feather cover between 37 and 48 weeks of age in the QTC compartments as compared to the control compartments. Separating males from females did not increase aggressive behaviour between the males in the male pen (Van Emous, 2010).

Figure 3 Quality Time® broiler breeder house. Males are penned in a separate area during feeding (central area on the picture) (Photo: R.A. van Emous).

As indicated in the current paragraph, management practices to improve mating behaviour and thus welfare of broiler breeders exist, with concurrently improved performance of the breeder flocks. These methods need to be further developed for practical application.

5.3 Quality of the feather cover

Quality of the feather cover, especially during the production period, is important with respect to the prevention of skin damage in females resulting from male mating behaviour and with respect to thermoregulation of the bird. However, quality of the feather cover has decreased during the past 10-20 years due to unknown reasons (Van Emous and De Jong, 2013). Figure 4 shows examples of a deteriorated feather cover in females in the second half of the production period.

Although not performing mutilations in broiler breeders implies an increased risk for a deteriorated feather cover (and skin damage), as indicated earlier, non-beak trimming of males and females did not seem to have a negative effect on the feather cover of both males and females (De Jong et al., 2013; De Jong, Gunnink and Van Emous, 2016, pers. comm.). Reduction of the stocking density improved the quality of the feather cover both in the rearing and in the production periods, although in general females at both standard and reduced stocking density still had a deteriorated feather cover at the end of the production period (De Jong et al., 2011).

Van Emous et al. (2014, 2015a) studied the effect of diets with different protein content in the rearing period and different energy content in the production period on the quality of the feather cover. They concluded that a low daily protein intake during the rearing period and during first phase of the production period resulted in an inferior feather cover as compared to diets with medium or high protein content at these ages (see Van Emous et al., 2014, 2015a for diet composition tables). This indicates that specific amino acids levels for feather development were deficient. The authors also suggested

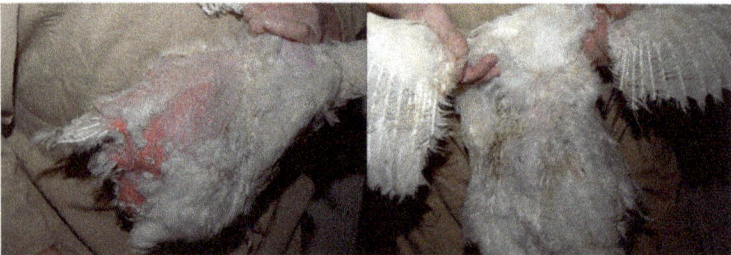

Figure 4 Examples of a deteriorated feather cover in female broiler breeders in the second half of the production period (Photo: R.A. van Emous).

that a low daily protein intake between 2 and 6 weeks of age showed a more pronounced effect on feather cover than a low daily protein intake between 6 and 15 weeks of age.

6 Environmental enrichment

It is well known that chickens prefer a high and safe resting place to roost at night-time, due to the motivation to protect themselves from predators. Usually in the rearing period, raised slatted areas ($\sim$1 m^2/1000 birds) are provided from six weeks onwards to train the birds to go to the higher level, to prepare them to easily enter the nesting area during the production period. However, it is yet unclear if raised slatted areas might function as high and safe resting places or if breeders prefer perches. In addition, how these perches should be provided, including material and shape, is yet unclear. Most studies on perch design and perch use are carried out in laying hens, but it needs to be studied if results from the studies regarding laying hens are also valid for broiler breeders. A Swedish study, comparing wooden A-shaped perches with platforms, found a high incidence of breast blisters in the broiler breeders on perches which was not found in birds provided with platforms (Wachenfelt and Berndson, 2014). This could be due to the material and shape of the perches. In addition, they observed a preference of the birds to rest on the platforms, but this could have been influenced by birds experiencing pain due to the breast blisters and thus not using the perches (Wachenfelt and Berndson, 2014). Other studies only recorded that perches were used by broiler breeders, with increasing use with age (Hocking and Jones, 2006; Van Emous, 2016, pers. comm.). Further study is required to find whether perches or platforms are preferred by broiler breeders and which design should be applied when using perches. It is known from the studies regarding laying hens that a high proportion of birds in non-cage systems may suffer from keel bone fractures and deformities, caused by collisions with housing structures or prolonged pressure on the keel bone during perching (Heerkens et al., 2016). It is not known if broiler breeders suffer from keel bone fractures and keel bone deviations due to the design of perches and platforms.

Hocking and Jones (2006) studied if environmental enrichment, by providing plastic-coated bales of wood shavings or bunches of string, could decrease aggressive pecking and feather damage in broiler breeders in the rearing period. It turned out that plastic-coated bales of wood shavings, but not strings, were attractive for the breeders. However, neither behavioural changes nor improved feather cover were observed in the groups where enrichment was provided as compared to the control groups without enrichment.

Vertically placed cover panels in the litter area in the production house can be used to control excessive mating problems in commercial farms

(Estevez, 1999) and thus may improve female broiler breeder welfare during the production period. In addition, cover panels improved reproductive performance in broiler breeder flocks, probably by attracting females to the litter floor and reducing male-male competition for females and over-mating (Leone and Estevez, 2008).

In general, environmental enrichment is not commonly applied in broiler breeder houses. Because environmental enrichment may improve the behavioural opportunities and reduce problematic behaviours such as injurious pecking (Estevez, 2009), further research to find appropriate enrichment for broiler breeders both in the rearing and in the production periods is needed.

7 Vaccinations

Maintaining flock health at a high level is an important issue during the rearing and the production period. Farmers usually apply a strict hygiene regime and disease prevention is an important issue with respect to flock management (EFSA, 2010). The applied vaccination schedule is dependent on country, region, individual farmers or advisers and integrations. However, during the rearing period broiler breeders can be subjected to a series of vaccinations starting at the hatchery or even *in ovo*. Examples are Marek's disease and infectious bronchitis (IB) vaccination at the hatchery, and vaccinations against Newcastle disease, IB, infectious laryngotracheitis (ILT), infectious rhinotracheitis, Gumboro, *Salmonella* and *Escherichia coli* at various ages in the rearing period. In the production period, broiler breeders are subjected to less frequent vaccinations (e.g. against IB), although this is also farmer or integration dependent. Vaccinations can be provided in various forms, such as injections, spray/aerosol, via drinking water and eye drops. Despite the need to prevent diseases in the breeder flock, (multiple) vaccinations are a challenge for the birds, resulting in depressed feeding (and thus growth) or local inflammation (Steentjes, 2016, pers. comm.).

8 Transgenerational effects

A relatively unknown area is how conditions during the life of broiler breeders affect the welfare and technical performance of the progeny. It is well known that age of the breeders is related to egg quality and chick quality, and first week mortality is higher in broilers from young breeding flocks (Yassin et al., 2009). However, it is important to consider that other conditions, for example, stress or diseases, in the life of the breeders may also affect the performance and welfare of the progeny. From laying hen studies it is known that feather pecking and anxiety in the layer breeding flock is related to the prevalence of feather pecking and anxiety in the progeny, although these effects were dependent on the genotype of the birds. The underlying mechanism may act through the deposition of corticosterone in the yolk of the eggs (de Haas et al., 2014).

Van der Waaij et al. (2011) studied the progeny of 60-week-old female broiler breeders that were either fed the recommended amount of feed or fed *ad libitum* during a 5-week period. The offspring of breeders fed *ad libitum* was heavier than the offspring of restricted fed breeders. In addition, female offspring of restricted fed breeders were lighter at hatching, but were heavier and had more abdominal fat at six weeks of age than female offspring of the *ad libitum* fed breeders. These results suggested a possible transgenerational effect of the feeding regime in breeders on the broilers. However, more research is needed to further explore this relationship, as reviewed by Berghof et al. (2013), who focused on possible transgenerational effects of stress in breeders on innate immunity in broiler chickens. They suggested that the mismatch between the breeder and broiler environment (e.g. by diet composition or microbiota) may negatively affect the innate immunity in the broiler chickens, but that more research is needed to further unravel the underlying mechanisms.

With respect to the effect of feed composition in the production phase of the breeders and its influence on the progeny, there is some evidence that low-density diets (21% lower nutrient density as compared to the standard diet) positively affect egg composition (higher egg white proportion) and hatchability, especially in young breeders (Enting et al., 2007b). The same authors found that these low-density broiler breeder diets improved growth rate of the progeny, reduced broiler mortality and either reduced or increased immune responses, depending on the breeder's age and egg weight (Enting et al., 2007a). It is not clear if these effects are due to transmission of nutrients from the breeder to the egg and the progeny, or to reduced stress in the breeders, or both.

Van Emous et al. (2015b) also studied the combined effects of low protein diets and different growth patterns in the rearing period on offspring performance. In general, they only found marginal effects of the different protein contents on progeny performance. However, male broilers of breeders fed the low protein diet during the rearing period had higher breast meat yield as compared to broilers of broiler breeders fed the medium or high protein diet – see Van Emous et al. (2015b) for diet composition tables. The effects of different daily protein intake of breeders during the rearing and the production period on progeny performance were observed in a consecutive experiment (van Emous et al., 2015a). A lower mortality and improved technical performance were observed in the progeny of 53-week-old breeders fed a low daily protein intake during the second phase of the laying period.

To conclude, there is evidence for transgenerational effects of management of broiler breeders and their progeny. This area needs further exploration to improve welfare and performance of both breeders and broilers.

9   Concluding remarks

The majority of the broiler breeders worldwide are parent stock of standard, fast growing broiler chickens. Because of the genetic selection for fast growth in the progeny, broiler breeders need to be subjected to restricted feeding programmes, especially during the rearing period. This is, despite a number of studies focusing on the methods to alleviate the negative effects of feed restriction on the welfare, still one of the most important welfare issues in broiler breeders. However, using feeds with higher insoluble fibre contents or lower protein content in the rearing period as compared to the standard commercial diets seems promising in terms of alleviating the effect of feed restriction on welfare. Another alternative is the use of slower growing or dwarf female broiler breeders that do not need to be fed restricted. Apart from feed restriction, other areas that need more attention in research are the relationship between management in the production period and mating behaviour, provision of environmental enrichment and how breeder management affects offspring performance.

10   Where to look for further information

Further information on housing and management can be found in management guides of the breeding companies that are updated on a regular basis (Aviagen, 2013; Cobb, n.d., Hubbard, n.d.) and textbooks (Leeson and Summers, 2000). Specific welfare issues have been discussed in various review papers (e.g. Decuypere et al., 2006, 2010; De Jong and Guemené, 2011) and the EFSA Scientific Opinion on welfare aspects of the management and housing of the grandparent and parent stocks raised and kept for breeding purposes (EFSA, 2010).

11   References

Aviagen. 2013. Ross parent stock management handbook. http://en.aviagen.com/assets/Tech_Center/Ross_PS/Ross_PS_Handbook_2013_i-r1.pdf (Accessed date 24 March 2016).

Berghof, T. V. L., H. K. Parmentier and A. Lammers. 2013. Transgenerational epigenetic effects on innate immunity in broilers: An underestimated field to be explored? *Poultry Science* 92:2904-13.

Bruggeman, V., O. Onagbesan, E. D'Hondt, N. Buys, M. Safi, D. Vanmontfort, L. Berghman, F. Vandesande and E. Decuypere. 1999. Effects of timing and duration of feed restriction during rearing on reproductive characteristics in broiler breeder females. *Poultry Science* 78:1424-34.

Buckley, L. A., L. M. McMillan, V. Sandilands, B. J. Tolkamp, P. M. Hocking and R. B. D'Eath. 2011a. Too hungry to learn? Hungry broiler breeders fail to learn a Y-maze food quantity discrimination task. *Animal Welfare* 20:469-81.

Buckley, L. A., V. Sandilands, P. M. Hocking, B. J. Tolkamp and R. B. D'Eath. 2015. Feed-restricted broiler breeders: State-dependent learning as a novel welfare assessment tool to evaluate their hunger state? *Applied Animal Behaviour Science* 165:124-32.

Buckley, L. A., V. Sandilands, B. J. Tolkamp and R. B. D'Eath. 2011b. Quantifying hungry broiler breeder dietary preferences using a closed economy T-maze task. *Applied Animal Behaviour Science* 133:216-27.

Casanovas, P. 2000. Management techniques to improve male mating activity and compensate for the age-related decline in broiler breeder fertility: Intra-spiking. *Cobb Technical News* 7:1-8.

Cobb, undated. Cobb500FF Breeder Mangement Supplement. http://www.cobb-vantress.com/docs/default-source/cobb-500-guides/cobb500ff-breeder-mangement-supplement---(english)3CA513F5C35A.pdf (Accessed date 24 March 2016).

D'Eath, R. B., B. J. Tolkamp, I. Kyriazakis and A. B. Lawrence. 2009. 'Freedom from hunger' and preventing obesity: the animal welfare implications of reducing food quantity or quality. *Animal Behaviour* 77:275-88.

Dawkins, M. S. and R. Layton. 2012. Breeding for better welfare: genetic goals for broiler chickens and their parents. *Animal Welfare* 21:147-55.

de Beer, M. and C. N. Coon. 2007. The effect of different feed restriction programs on reproductive performance, efficiency, frame size, and uniformity in broiler breeder hens. *Poultry Science* 86:1927-39.

Decuypere, E., V. Bruggeman, N. Li Everaert, R. Yue Boonen, J. De Tavernier, S. Janssens and N. Buys. 2010. The broiler breeder paradox: ethical, genetic and physiological perspectives, and suggestions for solutions. *British Poultry Science* 51(5): 569-79.

Decuypere, E., P.M. Hocking, K. Tona, O. Onagbesan, V. Bruggeman, E.K.M. Jones, S. Cassy, N. Rideau, S. Metayer, Y. Jego, J. Putterflam, S. Tesseraud, A. Collin, M. Duclos, J. J. Trevidy and J. Williams. 2006. Broiler breeder paradox: a project report. *World's Poultry Science Journal* 62:443-53.

de Haas, E. N., J. E. Bolhuis, B. Kemp, T. G. G. Groothuis and T. B. Rodenburg. 2014. Parents and early life environment affect behavioral development of laying hen chickens. *Plos ONE* 9. doi 10.1371/journal.pone.0090577.

De Jong, I. C., H. Enting, S. Van Voorst, E. W. Ruesink and H. J. Blokhuis. 2005a. Do low density diets improve broiler breeder welfare during rearing and laying? *Poultry Science* 84:194-203.

De Jong, I. C., M. Fillerup and H. J. Blokhuis. 2005b. Effect of scattered feeding and feeding twice a day during rearing on parameters of hunger and frustration in broiler breeders. *Applied Animal Behaviour Science* 92:61-76.

De Jong, I. C. and D. Guemene. 2011. Major welfare issues in broiler breeders. *World's Poultry Science Journal* 67:73-81.

De Jong, I. C., H. Gunnink and R. A. Van Emous. 2013. Monitoring van onbehandelde vleeskuikenmoederdieren op een praktijkbedrijf. 21 pp Wageningen UR Livestock Research, Lelystad, Report 716.

De Jong, I. C. and R. B. Jones. 2006. Feed restriction and welfare in domestic birds. In V. Bels (ed.), *Feeding in Domestic Vertebrates*, pp. 120-35. CABI publishing, Wallingford, UK.

De Jong, I. C., A. Lourens, H. Gunnink and R. A. Van Emous. 2011. Effect of stocking density on (the development of) sexual behaviour and technical performance in broiler breeders. 61 p. Wageningen UR Livestock Research, Lelystad, Report 457.

De Jong, I. C. and M. Swalander. 2013. Housing and management of broiler breeders and turkey breeders. In V. Sandilands and P. M. Hocking (eds), *Alternative Systems for Poultry: Health, Welfare and Productivity*, pp. 225-49. CABI publishing, Wallingford, UK.

De Jong, I. C., S. Van Voorst and H. J. Blokhuis. 2003. Parameters for quantification of hunger in broiler breeders. *Physiology & Behavior* 78:773-83.

De Jong, I. C., S. Van Voorst, D. A. Ehlhardt and H. J. Blokhuis. 2002. Effects of restricted feeding on physiological stress parameters in growing broiler breeders. *British Poultry Science* 43:157-68.

De Jong, I. C., M. Wolthuis-Fillerup and R. A. Van Emous. 2009. Development of sexual behaviour in commercially-housed broiler breeders after mixing. *British Poultry Science* 50:151-60.

Dixon, L. M., S. Brocklehurst, V. Sandilands, M. Bateson, B. J. Tolkamp and R. B. D'Eath. 2014. Measuring motivation for appetitive behaviour: food-restricted broiler breeder chickens cross a water barrier to forage in an area of wood shavings without food. *Plos ONE* 9. doi 10.1371/journal.pone.0102322.

EFSA. 2010. Scientific Opinion on welfare aspects of the management and housing of the grand-parent and parent stocks raised and kept for breeding purposes. *The EFSA Journal* 8:81. doi doi:10.2903/j.efsa.2010.1667.

Enting, H., W. J. A. Boersma, J. B. W. J. Cornelissen, S. C. L. van Winden, M. W. A. Verstegen and P. J. van der Aar. 2007a. The effect of low-density broiler breeder diets on performance and immune status of their offspring. *Poultry Science* 86:282-90.

Enting, H., T. A. M. Kruip, M. W. A. Verstegen and P. J. van der Aar. 2007b. The effect of low-density diets on broiler breeder performance during the laying period and on embryonic development of their offspring. *Poultry Science* 86:850-6.

Estevez, I. 1999. Cover panels for chickens: a cheap tool that can help you. In *Poultry Perspectives*, pp. 4-6.

Estevez, I. 2009. Behaviour and environmental enrichment in broiler breeders. In *Biology of Breeding Poultry*, pp. 261-83. CABI publishing, Wallingford, UK.

Fiks-van Niekerk, T. G. C. M., I. C. de Jong, T. Veldkamp, M. M. Van Krimpen and R. A. Van Emous. 2009. Mutliations in poultry. 46 p. Wageningen UR Livestock Research, Lelystad, Report 255 .

Fontana, W. A., A. D. Weaver and H. P. Van Krey. 1992. intermittend periods of infertility identified in naturally mated broiler breeder hens. *Journal of Applied Poultry Research* 1:190-3.

Gentle, M. J. and D. E. F. McKeegan. 2007. Evaluation of the effects of infrared beak trimming in broiler breeder chicks. *Veterinary Record* 160:145-8.

Heck, A., O. Onagbesan, K. Tona, S. Metayer, J. Putterflam, Y. Jego, J. J. Trevidy, E. Decuypere, J. Williams, M. Picard and V. Bruggeman. 2004. Effects of ad libitum feeding on performance of different strains of broiler breeders. *British Poultry Science* 45:695-703.

Heerkens, J. L. T., E. Delezie, T. B. Rodenburg, I. Kempen, J. Zoons, B. Ampe and F.A.M. Tuyttens. 2016. Risk factors associated with keel bone and foot pad disorders in laying hens housed in aviary systems. *Poultry Science* 95: 482-8.

Hiemstra, S. J. and J. Ten Napel. 2013. Study of the impact of genetic selection on the welfare of chickens bred and kept for meat production. 118 p. http://ec.europa.eu/ food/animals/docs/aw_practice_farm_broilers_653020_final-report_en.pdf.

Hocking, P. M. 1993. Welfare of broiler breeder and layer females subjected to food and water control during rearing: quantifying the degree of restriction. *British Poultry Science* 343:53-64.

Hocking, P. M. and R. Bernard. 1993. Evaluation of putative appetite suppressants in the domestic fowl. *British Poultry Science* 34:393-404.

Hocking, P. M. and R. Bernard. 2000. Effects of the age of male and female broiler breeders on sexual behaviour, fertility and hatchability of eggs. *British Poultry Science* 41:370-6.

Hocking, P. M. and E. K. M. Jones. 2006. On-farm assessment of environmental enrichment for broiler breeders. *British Poultry Science* 47:418-25.

Hocking, P. M., M. H. Maxwell and M. A. Mitchell. 1993. Welfare assessment of broiler breeder and layer females subjected to food restriction and limited access to water during rearing. *British Poultry Science* 34:443-58.

Hocking, P. M., M. H. Maxwell and M. A. Mitchell. 1996. Relationships between the degree of food restriction and welfare indices in broiler breeder females. *British Poultry Science* 37:263-78.

Hocking, P. M., M. H. Maxwell, G. W. Robertson and M. A. Mitchell. 2002. Welfare assessment of broiler breeders that are food restricted after peak of lay. *British Poultry Science* 43:5-15.

Hocking, P. M., D. Waddington, M. A. Walker and A. B. Gilbert. 1989. Control of the development of the ovarian follicular hierarchy in broiler breeder pullets by food restriction during rearing. *British Poultry Science* 30:161-74.

Hocking, P. M., V. Zaczek, E. K. M. Jones and M. G. McLeod. 2004. Different concentrations and sources of dietary fibre may improve the welfare of female broiler breeders. *British Poultry Science* 45:9-19.

Horne, P. L. M. and N. Bondt. 2014. Competitiveness of the EU poultry meat sector. 46 p. LEI Wageningen UR, Wageningen, Report nr LEI 2014-038.

Hubbard, undated. Guide parent stock. http://www.hubbardbreeders.com/media/ guide__ps__classic__08__01__2016_imp__051137500_1634_22012016.pdf (Accessed date 24 March 2016).

Jones, E. K. M. and N. B. Prescott. 2000. Visual cues used in the choice of mate by fowl and their potential importance for the breeder industry. *World's Poultry Science Journal* 56:127-38.

Jones, E. K. M., N. B. Prescott, P. Cook, R. P. White and C. M. Wathes. 2001. Ultraviolet light and mating behaviour in domestic broiler breeders. *British Poultry Science* 42:23-32.

Jones, E. K. M., V. Zaczek, M. McLeod and P. M. Hocking. 2004. Genotype, dietary manipulation and food allocation affect indices of welfare in broiler breeders. *British Poultry Science* 45:725-37.

Leeson, S. and J. D. Summers. 2000. *Broiler Breeder Production*. Unviersity Books, Guelph, Canada.

Leone, E. H. and I. Estevez. 2008. Economic and welfare benefits of environmental enrichment for broiler breeders. *Poultry Science* 87:14-21.

Leone, E. H., I. Estevez and M. C. Christman. 2007. Environmental complexity and group size: Immediate effects on use of space by domestic fowl. *Applied Animal Behaviour Science* 102:39-52.

Mench, J. A. 2002. Broiler breeders: feed restriction and welfare. *World's Poultry Science Journal* 58:23-30.

Millman, S. T. and I. J. H. Duncan. 2000a. Effect of male-to-male aggressiveness and feed-restriction during rearing on sexual behaviour and aggressiveness towards females by male domestic fowl. *Applied Animal Behaviour Science* 70:63-82.

Millman, S. T. and I. J. H. Duncan. 2000b. Strain differences in aggressiveness of male domestic fowl in response to a male model. *Applied Animal Behaviour Science* 66:217-33.

Millman, S. T., I. J. H. Duncan and T. M. Widowski. 2000. Male broiler breeder fowl display high levels of aggression toward females. *Poultry Science* 79:1233-41.

Moradi, S., M. Zaghari, M. Shivazad, R. Osfoori and M. Mardi. 2013. Response of female broiler breeders to qualitative feed restriction with inclusion of soluble and insoluble fiber sources. *Journal of Applied Poultry Research* 22:370-81.

Morrissey, K. L. H., T. Widowski, S. Leeson, V. Sandilands, A. Arnone and S. Torrey. 2014a. The effect of dietary alterations during rearing on feather condition in broiler breeder females. *Poultry Science* 93:1636-43.

Morrissey, K. L. H., T. Widowski, S. Leeson, V. Sandilands, A. Arnone and S. Torrey. 2014b. The effect of dietary alterations during rearing on growth, productivity, and behavior in broiler breeder females. *Poultry Science* 93:285-95.

Nielsen, B. L., K. Thodberg, J. Malmkvist and S. Steenfeldt. 2011. Proportion of insoluble fibre in the diet affects behaviour and hunger in broiler breeders growing at similar rates. *Animal* 5:1247-58.

Renema, R. A., M. E. Rustad and F. E. Robinson. 2007. Implications of changes to commercial broiler and broiler breeder body weight targets over the past 30 years. *World's Poultry Science Journal* 63:457-72.

Sandilands, V., B. J. Tolkamp and I. Kyriazakis. 2005. Behaviour of food restricted broilers during rearing and lay - effects of an alternative feeding method. *Physiology & Behavior* 85:115-23.

Sandilands, V., B. J. Tolkamp, C. J. Savory and I. Kyriazakis. 2006. Behaviour and welfare of broiler breeders fed qualitatively restricted diets during rearing: Are there viable alternatives to quantitative restriction? *Applied Animal Behaviour Science* 96:53-67.

Savory, C. J. and J.-M. Lariviere. 2000. Effects of qualitative and quantitative food restriction treatments on feeding motivational state and general activity level of growing broiler breeders. *Applied Animal Behaviour Science* 69:135-47.

Savory, C. J. and J. S. Mann. 1997. Is there a role for corticosterone in expression of abnormal behaviour in restricted-fed fowls? *Physiology & Behavior* 62:7-13.

Savory, C. J. and K. Maros. 1993. Influence of degree of food restriction, age and time of day on behaviour of broiler breeder chickens. *Behavioural Processes* 29:179-90.

Savory, C. J., K. Maros and S. M. Rutter. 1993. Assessment of hunger in growing broiler breeders in relation to a commercial restricted feeding programme. *Animal Welfare* 2:131-52.

Savory, C. J., E. Seawright and A. Watson. 1992. Stereotyped behaviour in broiler breeders in relation to husbandry and opiod receptor blockade. *Applied Animal Behaviour Science* 32:349-60.

Published by Burleigh Dodds Science Publishing Limited, 2017.

Savory, C. J., F. A. M. Tuyttens and J. S. Mann. 1996. Temporal patterning of oral stereotypies in restricted-fed fowls: 2. influece of meal frequency and meal size. *International Journal of Comparative Psychology* 9:140-58.

Tolkamp, B. J., V. Sandilands and I. Kyriazakis. 2005. Effects of qualitative feed restriction during rearing on the performance of broiler breeders during rearing and lay. *Poultry Science* 84:1286-93.

van der Waaij, E. H., H. van den Brand, J. A. M. van Arendonk and B. Kemp. 2011. Effect of match or mismatch of maternal-offspring nutritional environment on the development of offspring in broiler chickens. *Animal* 5:741-8.

Van Emous, R. A. and I. C. De Jong. 2013. Promising management measures to solve the major welfare problems in broiler breeders. *Proceedings 2nd International Poultry Meat Congress Antalya* (Turkey).

Van Emous, R. A. 2010. Quality Time®; an innovative housing concept for broiler breeders. In *Proceedings of of the 2nd International symposium*. Highlights in nutrition and welfare in poultry production. Wageningen, The Netherlands, pp. 37-44.

van Emous, R. A., R. Kwakkel, M. van Krimpen and W. Hendriks. 2014. Effects of growth pattern and dietary protein level during rearing on feed intake, eating time, eating rate, behavior, plasma corticosterone concentration, and feather cover in broiler breeder females during the rearing and laying period. *Applied Animal Behaviour Science* 150:44-54.

van Emous, R. A., R. Kwakkel, M. van Krimpen and W. Hendriks. 2015a. Effects of different dietary protein levels during rearing and different dietary energy levels during lay on behaviour and feather cover in broiler breeder females. *Applied Animal Behaviour Science* 168:45-55.

van Emous, R. A., R. P. Kwakkel, M. M. van Krimpen, H. van den Brand and W. H. Hendriks. 2015b. Effects of growth patterns and dietary protein levels during rearing of broiler breeders on fertility, hatchability, embryonic mortality, and offspring performance. *Poultry Science* 94:681-91.

Wachenfelt, E. and E. Berndson. 2014. Usge of perches in meat fowlsSveriges lantbruksuniversitet Fakulteten för landskapsarkitektur, trädgårds- och växtproduk tionsvetenskap.

Yassin, H., A. G. J. Velthuis, M. Boerjan and J. van Riel. 2009. Field study on broilers' first-week mortality. *Poultry Science* 88:798-804.

Zuidhof, M. J., D. E. Holm, R. A. Renema, M. A. Jalal and F. E. Robinson. 2015. Effects of broiler breeder management on pullet body weight and carcass uniformity. *Poultry Science* 94:1389-97.

Zuidhof, M. J., R. A. Renema and F. E. Robinson. 2007. Reproductive efficiency and metabolism of female broiler breeders as affected by genotype, feed allocation, and age at photostimulation. 3. Reproductive efficiency. *Poultry Science* 86:2278-86.

Zuidhof, M. J., B. L. Schneider, V. L. Carney, D. R. Korver and F. E. Robinson. 2014. Growth, efficiency, and yield of commercial broilers from 1957, 1978, and 2005. *Poultry Science* 93:2970-82.

Chapter 4

Welfare issues affecting broiler breeders

Anja Brinch Riber, Aarhus University, Denmark

1   Introduction

The demand for broiler meat has been growing for decades, and broiler meat represents the major animal protein source in many countries around the world (OECD/FAO, 2018). One key factor for this popularity is the low cost of broiler meat compared to other meat types. To meet the demand for cheap and increasing amounts of broiler meat, the broiler industry has consistently selected for fast growth, a body composition with a high yield of breast meat and increased feed efficiency.

The structure of the broiler industry is divided into a primary breeding sector and a production sector (Fig. 1; Pollock, 1999). In the primary breeding sector, pedigree stock (i.e. pure lines, PL) produces progeny that goes on to the great grandparent (GGP) generation, followed by the grandparent (GP) generation. The grandparents produce the parent stock (PS) which passes to the production sector. Parent stock males and females are different breeds. The progeny of the parent stock is the final product, that is, the broiler chicken produced for meat. Worldwide, the parent stock consisted of around 410 million birds (Elfick, 2010) in 2009 with Europe being responsible for an estimated 60 million (Hiemstra and Napel, 2013). The global market for the parent stock in 2018 was estimated to have increased to 530-539 million birds (Elfick, pers. comm.). This chapter regards the parent stock, also referred to as the broiler breeders.

Around the world, the far majority of broiler breeders are of the fast growing genotypes such as Ross 308 and Cobb500. The growth rate of the offspring of these breeders is presently around 63 g per day, averaged over the lifespan of birds slaughtered at a target weight of 2.2 kg (Aviagen, 2019a; Cobb, 2018).

http://dx.doi.org/10.19103/AS.2020.0078.18

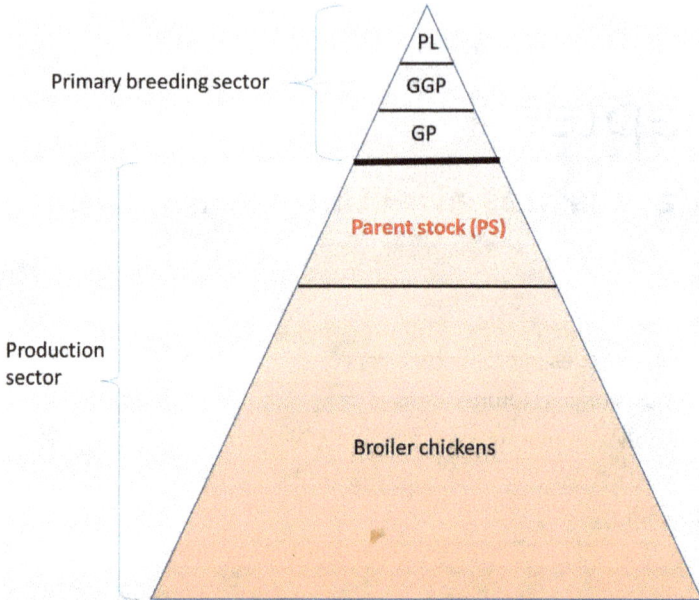

Figure 1 The structure of the broiler industry, showing the pure lines (PL), great grandparents (GGP) and grandparents (GP) belonging to the primary breeding sector, and the parent stock (PS) and the end product, the broiler chickens, belonging to the production sector. ©Anja Brinch Riber, Aarhus University, Denmark.

Due to the continued selection for fast growth, the growth rate is expected to keep increasing. Different genotypes, usually with a slower growth rate, are normally used for the middle segment or organic production, but presently these genotypes constitute a small fraction of the worldwide production of broiler meat. Of the approximately 60 million broiler breeders in Europe, an estimated 4.5 million are alternative genotypes (Horne, 2018).

1.1 Housing conditions of broiler breeders

Conventional broiler breeders are reared in single-sex flocks. In Europe, broiler breeders are mainly housed in fully littered floor-house systems, but rearing in cages is practised in other parts of the world. Depending on the climate and part of the world, the barn may either be a closed-house system or an open-sided system. Most closed-house systems are lit with artificial light at a light intensity down to approximately 5 lux (EFSA, 2010). The environment is typically barren, for example, there are often no perches or raised slats available (Fig. 2a).

At around 18-23 weeks of age, the birds are moved to the production farm where the two sexes are mixed with a ratio of around one male per 10-14 hens in partly littered floor-house systems (Aviagen, 2019b). Exceptions exist

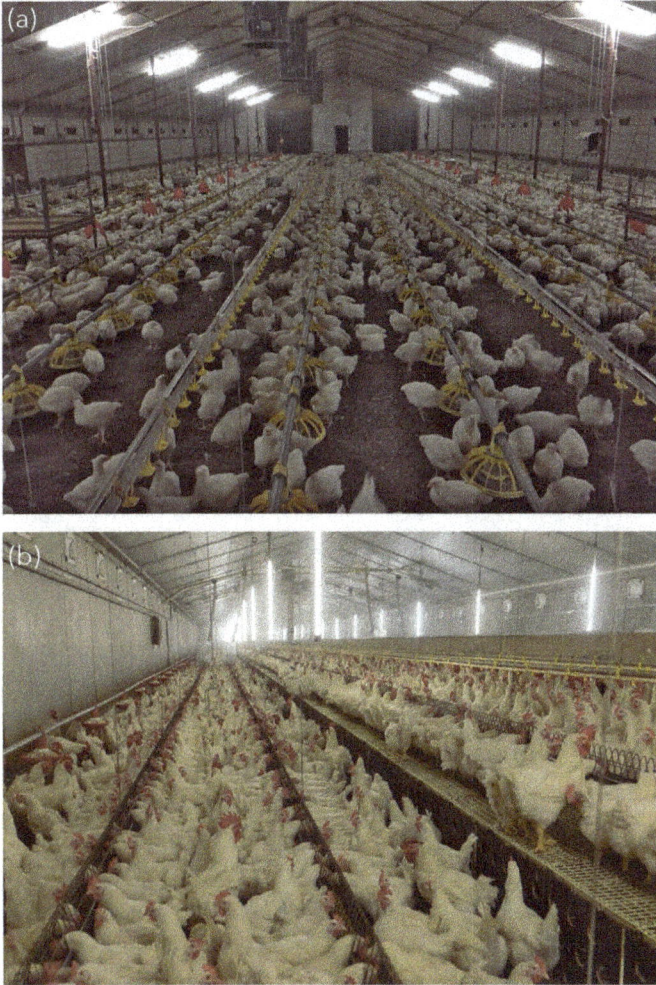

Figure 2 Examples of (a) a rearing farm using a fully littered floor-house system and (b) a production farm using a partly littered floor-house system. In the latter, the feeders for the males are seen to the left in the photo. The three feed lines covered with grids are accessible by the females. A water line with yellow water nipples are visible to the right in the photo and behind those, the nest boxes can be seen. ©Rick van Emous, Wageningen Livestock Research, The Netherlands.

with the sexes being housed separately in cages and artificial insemination being practised. From day-old, the surplus males are typically reared for meat in broiler production systems. Like during the rearing period, the barn may be either a closed-house system or an open-sided system. A light intensity of 30-60 lux is recommended (Aviagen, 2019b). When housed in floor-house systems, both drinking lines and nest boxes are typically placed on raised slats (Fig. 2b). The production period usually lasts until 60-65 weeks of age. To increase the fertility, a procedure called spiking is often practised where older males (>40 weeks of age) are replaced by younger males (Ordas et al., 2015).

1.2 Growth potential and feed restriction of broiler breeders

Genetic selection for growth parameters in broiler breeders has caused an increased appetite due to modulation of mechanisms of hunger regulation (Denbow, 1989; Siegel and Wolford, 2003). As a result, ad libitum access to feed will result in obesity and consequently in serious health and reproduction problems during the breeding period (Heck et al., 2004; Renema and Robinson, 2004). This negative correlation between the fast growth and reproductive success has been identified as the broiler breeder paradox (Decuypere et al., 2010).

To prevent health and reproductive problems, broiler breeders are feed restricted, except for the first 7-10 days of life. The level of feed restriction depends on age and sex of the birds. In terms of the amount of feed ingested, feed restriction is at its most severe level around age 10-16 weeks (Arrazola, 2018) where restrictively fed female broiler breeders are allocated down to four times less than ad libitum fed individuals will eat (de Jong et al., 2002; Savory et al., 1996). Of importance is also the level of feed restriction in terms of nutrient intake. It has been suggested that the feed restriction in terms of nutrient intake is most severe around age 5-7 weeks where female broiler breeders are allocated around 20% of the ad libitum nutrient intake (van Emous, unpublished data).

During the production period, feed restriction continues, but at a less restricted level for the female broiler breeders, with an allowance of 40-80% of the ad libitum intake depending on the age of the birds (Arrazola, 2018; Bruggeman et al., 1999). Males are typically fed less restrictively during rearing but more during the production period compared to females (Renema et al., 2007; EFSA, 2010). When the sexes are housed together during the production period, this is solved by using two kinds of feed troughs - feed troughs for males are hanged at a height where only the males can reach them, while the feed troughs for females are constructed in a way that makes the males unable to reach the feed because of their wider heads (Laughlin, 2009).

The alternative broiler breeder production differs markedly from the conventional, as other genotypes with lower growth potentials are used (see example shown in Fig. 3). Therefore, the level of feed restriction is not as severe and can even be omitted for some genotypes. One alternative type of broiler breeders is the dwarf lines where the females are affected by dwarfism (Decuypere et al., 2010). Dwarf broiler breeder females have a reduced growth, resulting in limited need or, for some lines, no need for feed restriction (Decuypere et al., 2010). As the dwarf gene is sex linked, the males will still

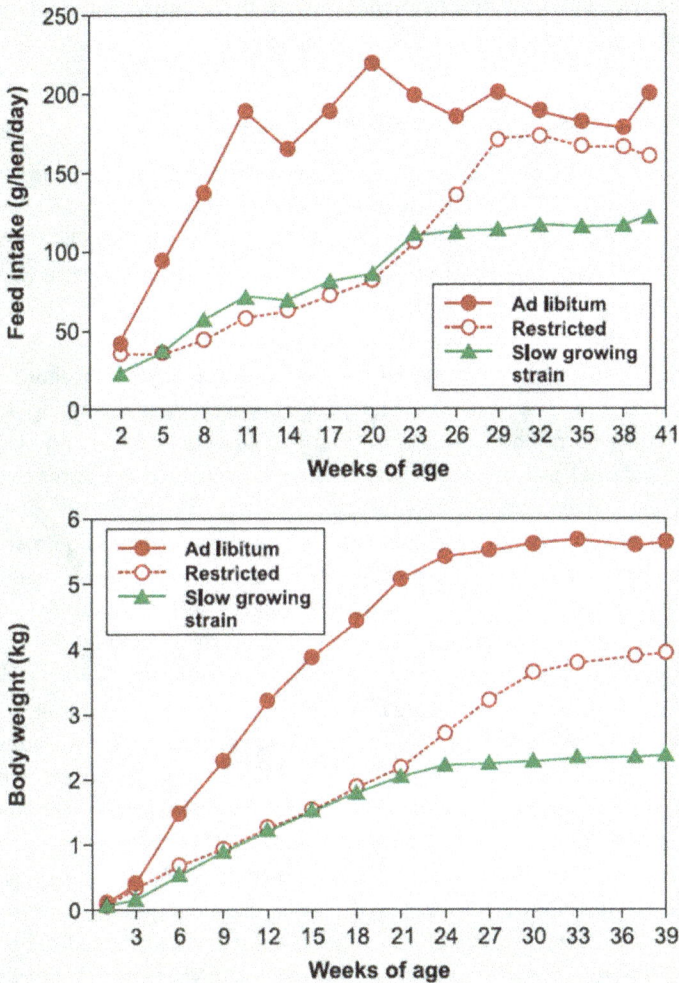

Figure 3 An example of the mean daily feed intake per broiler breeder female (top) and body weight (bottom) in a fast-growing genotype fed *ad libitum* (continuous red line) or restrictively (dashed red line) and in a slow-growing genotype (green line). ©William (2012).

need to be fed restrictively, but as the number of breeder males compared to breeder females is considerably lower, feed restriction will only be applied to a minor fraction of the broiler breeders.

In terms of welfare, the parent stock of fast-growing genotypes is facing significantly more welfare challenges due to the necessity of keeping them on a restricted diet to avoid obesity and the associated health deterioration and infertility. In addition, knowledge on welfare issues affecting organic broiler breeders is very sparse. The remaining part of this chapter will therefore focus on broiler breeders of the fast-growing genotypes.

2   Welfare issues in broiler breeders

The severe feed restriction introduces another paradox in broiler breeder production: the negative correlation between keeping the birds healthy and reproductive by the use of feed restriction and the welfare problems introduced by feed restricting the birds. Basic behavioural and physiological needs are commonly not fulfilled, and, as a result, the broiler breeders develop abnormal behaviour indicative of frustration and hunger (see review in D'Eath et al., 2009). Physiological stress responses in feed-restricted broiler breeders are also frequently found (D'Eath et al., 2009). The practise of limiting access to water and the barren housing environment, where elevated resting places and foraging material are scarce or absent resources, further contributes to the lack of satisfying behavioural and physiological needs.

Another major welfare issue in broiler breeders is aggression, particularly sexual aggression. Studies have reported on the loss of courtship behaviour, resulting in males chasing females, forced matings and increased levels of fear in females (Millman et al., 2000; Jones et al., 2001; de Jong et al., 2009). Linked to this is the mutilation of broiler breeders as a preventive measure to reduce the skin damage inflicted to flock mates. The beak of both sexes is typically trimmed, and, for the males, also the spurs and/or the outermost part of the toes facing backwards are commonly trimmed (EFSA, 2010). The mutilation of the males is to the benefit of the hens, but the procedures have negative consequences for the welfare of the males, introducing yet another broiler breeder paradox.

In contrast to the bulk of knowledge on welfare issues affecting laying hens and broilers, research conducted on the welfare challenges in broiler breeders is rather limited. There are probably multiple reasons for this, including limited accessibility to broiler breeder farms, limited awareness among NGOs, authorities and consumers of the welfare issues affecting broiler breeders and low priority when it comes to funding broiler breeder studies due to the relatively low number of individuals (as compared to laying hens or broilers). Anyway, the above-mentioned welfare issues in the broiler breeder industry are described in further detail in this section. Furthermore, methods of reducing the extent of the problems are discussed.

2.1 Unfulfilled behavioural and physiological needs

Physiological needs include the need for food, water and thermal comfort. The term 'behavioural needs' has been debated more, but, generally, a behavioural need is the need to perform a specific behaviour pattern regardless of the quality of the environment or whether the physiological needs, which the behaviour serves, are already fulfilled (Jensen and Toates, 1993). Thus, the animal's motivation to perform the behaviour plays a key role in whether the behavioural needs are perceived as satisfied. Animals that are highly motivated to perform a specific behaviour pattern may experience frustration and suffering if the performance of the behaviour is thwarted.

For conventional broiler breeders that are feed restricted and typically kept in barren environments, the behaviour pattern that is most commonly thwarted is feeding, including both the appetitive phase and the consummatory phase. Furthermore, in some parts of the world, resting in elevated positions is impeded during the rearing period due to the lack of perches or raised slats. With regard to physiological needs, broiler breeders experience metabolic hunger and may also suffer from the water restriction implemented. These welfare issues are discussed below.

2.1.1 Restrictive feeding and associated welfare problems

Being a motivated behaviour, feeding consists of two phases: the appetitive phase where the bird searches for food and gains food-related knowledge, that is, foraging, and the consummatory phase where the bird ingests food, that is, eating (Craig, 1918). Due to the feed restriction, the possibility of fulfilling the behavioural need for the consummatory phase of feeding is thwarted in broiler breeders. The daily ration of feed is quickly consumed, which may be as fast as in 4–16 min (Kostal et al., 1992). After ingestion of the daily ration, the broiler breeders are still highly motivated to continue feeding, partly because of hunger, and partly because of the unfulfilled behavioural need for carrying out feeding behaviour.

Thwarted possibility of performing the consummatory phase of feeding behaviour naturally causes an increase in time spent on the appetitive phase of feeding, that is, searching for food. Thus, as feed becomes restricted, more foraging behaviour in terms of pecking and scratching is performed, and also other oral activities may be increased. This has indeed been reported in several studies (Puterflam et al., 2006; Hocking et al., 1996, 2001, 2002; Savory and Maros, 1993; de Jong et al., 2003; Merlet et al., 2005). It may be questioned whether the increased foraging activities have any truly fulfilling effect on the behavioural need for feeding, particularly as the environment of broiler breeders is typically kept barren with limited access to foraging materials of high quantity and quality.

Certainly, feed-restricted broiler breeders show behaviour indicating frustration and hunger during both the rearing period and the production period. This involves increased general activity, in particular locomotion, indicating restlessness (Puterflam et al., 2006; de Jong et al., 2003). This comes at the expense of time spent on resting, eating and performing comfort behaviour (Puterflam et al., 2006; de Jong et al., 2003; Hocking et al., 2001, 2002) with the exception that at low levels of feed restriction, the broiler breeders may stand more, presumably to promote loss of heat resulting from the increased metabolism.

Furthermore, the feed-restricted broiler breeders often develop stereotypic pecking, that is, unvarying, repetitive pecking that has no obvious goal or function (Mason, 1991), directed towards objects in the environment (Puterflam et al., 2006; Nielsen et al., 2011). Stereotypic pecks may be directed towards the drinking nipples, resulting in both water spillage and excessive drinking (i.e. polydipsia, Hocking et al., 1993). In order to avoid water spillage and wet manure, which results in wet bedding and increased risk of contact dermatitis, the water consumption is controlled. This is partly done by reducing the water pressure, and partly by time limitations of access to the water. Few studies have investigated the effect on welfare of water restriction specifically. In general, water restriction is considered negative with respect to animal welfare, although one study showed that water restriction during rearing did not influence physiological welfare indicators in broiler breeders (Hocking et al., 1993).

The competition for feed results in an increased level of aggressive pecks, both among breeder males and among breeder females (Shea et al., 1990; Hocking and Jones, 2006; Hocking et al., 2005). Feather pecking, particularly of the tail feathers of conspecifics, is observed in broiler breeders (Morrissey et al., 2014; Girard et al., 2017a). It may have a stereotypic nature and take the form of 'tail sucking' (Morrissey et al., 2014; Nielsen et al., 2011; Girard et al., 2017a). Cannibalism is also reported, and studies show that alleviating hunger may reduce the risk of cannibalism, although not consistently (Hocking et al., 2004b; Nielsen et al., 2011). Although mating is thought to be the major cause of feather and skin damage in broiler breeders, feather pecking and cannibalistic pecks may also be important factors (de Jong and Guemene, 2011).

Physiologically, feed-restricted broiler breeders show signs of stress in terms of an increased relation between hetero granulocytes and lymphocytes (H/L) (Hocking et al., 1993, 1996; Bowling et al., 2018) and a higher level of plasma corticosterone (Hocking et al., 1993, 1996; de Jong et al., 2003). However, it remains debated whether this can be assigned to stress alone, as increased levels of plasma corticosterone may also occur due to metabolic reasons (de Jong et al., 2003; D'Eath et al., 2009).

2.1.2 The alleviation of hunger and fulfilment of behavioural needs in broiler breeders

Different feeding strategies have been used to alleviate the hunger felt by broiler breeders and to fulfil their behavioural need for feeding behaviour with varying success. Among these are qualitative feed restrictions, manipulating the number of daily meals, scattering of the feed by the use of spin feeders and appetite suppressants.

2.1.2.1 Qualitative feed restriction

The idea of qualitative feed restriction is to reduce the quality of the feed in terms of nutrient content by adding non-nutritious or poor-nutritious diluents to the standard feed. This can be one or more types of fibres, and examples of fibres used are sugar beet, out hulls, sunflower flour and even sawdust. When applying qualitative feed restriction, a larger ration of feed is needed to obtain the same nutrient intake, and the content in different parts of the gastrointestinal system of the birds increases (Steenfeldt and Nielsen, 2012; Hocking et al., 2004a).

Further increase of gut fill may be obtained by adding soluble types of fibres to the feed. Soluble fibres can absorb more water, thus increasing the intestinal content more than insoluble fibres (Hocking et al., 2004b). However, the use of soluble fibres in the feed has to be done with care, as too high levels of soluble fibres leads to a swelling of the feed in the intestines in prolonged periods after feeding, which may be painful to the birds (Nielsen et al., 2011). The higher consumption of water in broiler breeders fed a diet high on soluble fibres may also result in a deterioration of the bedding, which increases the risk of contact dermatitis, compromises resting comfort and reduces foraging and dust bathing opportunities.

By increasing the intestinal contents, the broiler breeders are expected to obtain an improved feeling of satiety. In addition to increased satiety, applying qualitative feed restriction increases the chance of meeting the behavioural need of the broiler breeders for performing feeding behaviour, as the time spent on feeding increases when a larger amount of feed is to be ingested (Zuidhof et al., 1995; Savory et al., 1996; de Jong et al., 2005a; Moradi et al., 2013).

Indeed, studies comparing broiler breeders fed restrictively, either qualitatively or quantitatively (standard), have found positive effects of the former feeding strategy on a range of welfare indicators. For instance, qualitative feed restriction results in less water intake and reduced time spent on drinking behaviour (Zuidhof et al., 1995; Savory et al., 1996; Hocking, 2006). Other studies have found fewer stereotypic pecks on objects (Savory et al., 1996; Hocking et al., 2004b; de Jong et al., 2005a; Nielsen et al., 2011) and less stereotypic tail feather pecking (Nielsen et al., 2011) in the qualitatively feed-restricted birds.

To assess hunger in qualitatively feed-restricted broiler breeders, de Jong et al. (2005a) used a test of compensatory food intake where the amount of feed eaten besides the ordinary daily allowance was examined. The authors found that the compensatory food intake was reduced at 12 weeks of age in broiler breeder pullets allocated low-density feed compared to standard feed. Likewise, Nielsen et al. (2011) showed that regardless of the type of fibres, the compensatory feed intake was lower in a test of broiler breeders fed a fibre-rich feed compared to standard feed. The compensatory food intake test is a validated method for quantifying the level of hunger, as it has been shown that the more feed restricted the broiler breeders have been, the higher feed intake they will have for an extended period (at least 4 days) when allowed *ad libitum* access to feed (de Jong et al., 2003).

In regard to the effect of qualitative feed restriction on physiological stress, contradicting results have been found concerning both the relation between hetero granulocytes and lymphocytes (Zuidhof et al., 1995; Savory et al., 1996; Hocking et al., 2004b; de Jong et al., 2005a; Hocking, 2006; Jones et al., 2004) and the level of plasma corticosterone (Savory et al., 1996; Hocking, 2006; Moradi et al., 2013). van Emous et al. (2014) explained it as a reflection of a general increase in the content of plasma corticosterone of broiler breeders due to an increased feed restriction during the last 30 years. As a result, the broiler breeders respond less to differences in feed allowance.

To summarise, the qualitative feed restriction appears to have a positive effect on behaviour and welfare, although dependent on the type and amount of the fibres used. However, common for the studies conducted up until now is that none of the fibre-rich diets examined can be characterised as being very effective in terms of alleviating hunger without negatively affecting other welfare aspects. Even though there are positive effects of qualitative feed restriction on satiety, it is uncertain whether it fully satisfies the broiler breeders' motivation for feeding behaviour. Besides, there is a risk that qualitatively feed-restricted broiler breeders continue to experience metabolic hunger (de Jong et al., 2005a).

2.1.2.2 Number of daily meals

Due to the feed restriction applied in the conventional broiler breeder industry, feeding programmes are always used. These programmes differ in the number of meals allocated daily (Fig. 4). Feeding once a day may be applied - either by choice or due to daily feeding being a requirement in the national legislation (e.g. UK, Sweden and Denmark). As an alternative, skip-a-day feeding programmes may be used (EFSA, 2010). It involves either one, two or three days (non-consecutive) without feed per week. On feeding days, the birds are fed a larger portion than if they were fed daily.

The effect on welfare of skip-a-day feeding programmes is, based on the few studies available, inconclusive. Lindholm et al. (2018) found signs of increased

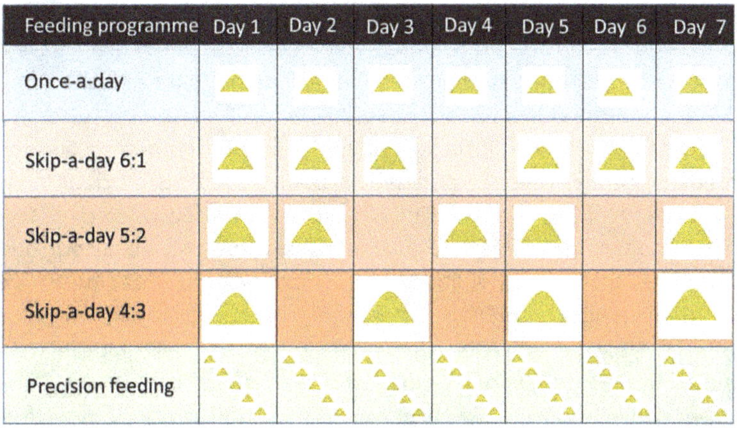

Feeding programme	Day 1	Day 2	Day 3	Day 4	Day 5	Day 6	Day 7
Once-a-day	▲	▲	▲	▲	▲	▲	▲
Skip-a-day 6:1	▲	▲	▲		▲	▲	▲
Skip-a-day 5:2	▲	▲		▲	▲		▲
Skip-a-day 4:3	▲		▲	▲			▲
Precision feeding	▲▲▲	▲▲▲	▲▲▲	▲▲▲	▲▲▲	▲▲▲	▲▲▲

Figure 4 Outline of the number and portions of meals during a week on the different feeding programmes. Portions not to scale. ©Anja Brinch Riber, Aarhus University, Denmark.

physiological stress in broiler breeder pullets experiencing 2 days a week without food compared to birds fed daily. This included elevated heterophil to lymphocyte ratios, increased adiposity and reduced muscle growth. On the other hand, the skip-a-day birds showed signs of lower anxiety before feeding times, which may be a result of the lower feed competition associated with larger portion sizes. Skip-a-day birds generally showed more interest in a novel object in the home pen, which indicates increased risk taking and reduced fear while fasting. No signs were found of the skip-a-day birds learning the feeding schedule, and this unpredictability may also increase stress.

In contrast, Skinner-Noble and Teeter (2009a cited in EFSA, 2010) found no signs of increased stress levels in skip-a-day birds compared to daily-fed birds, neither when using behavioural or physiological indicators. Furthermore, the plumage condition of broiler breeders has been found to be better during the early production period in skip-a-day birds compared to birds fed daily (Morrissey et al., 2014).

A new feeding system has recently been developed for broiler breeders: the precision feeding system (Zuidhof et al., 2017). In this system, small meals are provided to the individual bird multiple times each day. It works such that a bird can enter the system voluntarily at any time. The bird is automatically weighed, and if its body weight is below the target weight, access to feed is allowed for a limited period. When using the precision feeding system, the uniformity of the flock has been shown to increase (Zuidhof, 2018; Zuidhof et al., 2017).

Compared to skip-a-day feeding, broiler breeders with access to the precision feeding system show less restlessness behaviour, that is, more sitting and less standing and walking (Girard et al., 2017a). In addition, they perform less foraging and feather pecking behaviour (Girard et al., 2017a). However, increased aggression and more stereotypic pecking at the drinkers have been observed in precision-fed broiler breeders compared to those fed on the skip-a-day feeding programme (Girard et al., 2017a,b).

The effect of feeding twice daily, compared to once daily, on hunger and frustration in broiler breeder pullets has been examined by de Jong et al. (2005b). The result was unclear. The broiler breeders fed twice daily showed more locomotion, which may indicate increased foraging behaviour but also frustration because of an unfulfilled behavioural need of foraging. The other indicators of hunger examined (i.e. concentration of plasma-corticosterone, compensatory food intake and relation between plasma and glucose) were not influenced by whether the birds were fed once or twice daily.

Based on the existing knowledge, it is clear that the applied feeding programme in the broiler breeder industry is likely to affect the behaviour and welfare of the birds. However, deriving clear recommendations of which feeding programme to apply is hampered by the inconsistency of results from different studies of the same feeding programme and by the different directions in which welfare indicators are pointing when comparing different feeding programmes.

2.1.2.3 Scattering of feed in the bedding

Scattering of feed in the litter during rearing of broiler breeder pullets is a feeding method practised in some parts of the world, particularly Europe and Northern America (EFSA, 2010). The scattering of feed is commonly done by using the so-called spin feeders. Using this method, the feed is quickly and evenly distributed to all the birds within the flock, promoting uniform feed intake (Hocking, 2004). In addition, the feeding method stimulates foraging behaviour. de Jong et al. (2005b) studied the effect of scattering all the feed in the litter during rearing of broiler breeder pullets. Stereotypic pecking on objects was reduced, but no other indicators of hunger, that is, the concentration of plasma corticosterone, compensatory food intake and the relation between plasma and glucose, were influenced by whether the feed was scattered in the bedding or allocated in feed troughs.

2.1.2.4 Appetite suppressants

Appetite suppressants have been suggested as a method to alleviate the hunger felt by feed-restricted broiler breeders. The idea is that the use of these additives suppresses the appetite to a level where quantitative feed restriction is no longer necessary for the birds to remain on the target body weight. However, the method has only proved partly successful in improving the welfare of the birds (Sandilands et al., 2006; Savory et al., 1996). Concern has been expressed whether the reducing effect on appetite derives from discomfort felt by the birds (EFSA, 2010). Ethically this may not be acceptable to consumers or society as such.

2.1.3 Lack of elevated resting places and associated welfare problems

Elevated resting places, that is, raised slats or perches, are used for resting both during the day and at night. Indeed, broiler breeders will use elevated resting places when given access (Gebhardt-Henrich et al., 2017). Although slow-growing genotypes will use the elevated resting places more frequently, fast-growing genotypes also have a high use initially, yet this will reduce with age (Gebhardt-Henrich and Oester, 2014). Raised slats are preferred to perches by broiler breeders (Gebhardt-Henrich and Oester, 2014).

Thus, similar to other types of domestic fowl, broiler breeders are motivated to use elevated resting places. However, in some parts of the world, elevated resting places are not provided during the rearing period, that is, the first access to elevated structures occurs after transfer to the production farm. In addition to fulfilling a behavioural need, provision of elevated resting places also has positive effects on other welfare aspects. For instance, fear has been shown to be reduced in broiler breeders reared with access to perches (Brake et al., 1994). Furthermore, it is known from laying hens that elevated resting places also function as shelters during the day to avoid aggressive interactions with flock mates (Cordiner and Savory, 2001).

Providing elevated resting places during the rearing period prepares the broiler breeders for the more complex environment in the production houses. Usually, the nest boxes are elevated from the ground and often water nipples are placed above elevated slats, forcing both males and females to use the three-dimensional space. Rearing in houses with elevated resting places will promote the broiler breeders' use of the three-dimensional space in the production houses, as early access to three-dimensional structures improves the use later in life (Gunnarsson et al., 2000; Norman et al., 2018).

Negative effects in terms of keel bone damage have been associated with the use of perches in laying hens (Stratmann et al., 2015), but Gebhardt-Henrich et al. (2017) found no difference between adult broiler breeder females reared with or without perches. Provision of elevated resting places is considered to have an overall positive effect on behaviour and welfare of broiler breeders.

2.2 Aggression and associated welfare problems

Non-sexual aggression is common both among broiler breeder males and among broiler breeder females (Shea et al., 1990; Hocking and Jones, 2006;

Hocking et al., 2005). One explanation for this is the increased competition for feed. The level of non-sexual aggression has been reported to be highest at the time of the day where feed is allocated (Shea et al., 1990). Furthermore, feed-restricted broiler breeders display more non-sexual aggression at the feeder than do broiler breeders fed *ad libitum* (Mench, 1988). Frustration resulting from feed restriction could be another causal factor for the increased non-sexual aggression. With skip-a-day feeding, it has been shown that aggression is most pronounced on mornings where no feed is allocated (Mench, 1988), although this was not confirmed in a later study (Shea et al., 1990).

A high level of sexual aggression is also present in the broiler breeder production (Millman and Duncan, 2000; Millman et al., 2000). The sexual behaviour in broiler breeders lacks many of the elements of normal sexual behaviour, which both the red jungle fowl and laying hens exhibit, with courtship behaviour prior to mating being almost non-existing (Millman et al., 2000; Jones et al., 2001; de Jong et al., 2009). The mating behaviour of the males is described as being rough where they chase and peck the hens and force them to mate (Millman et al., 2000; Jones et al., 2001).

The rough mating behaviour induces fear and reduces welfare in broiler breeder females (de Jong et al., 2009; Millman et al., 2000; Leone and Estevez, 2008). The plumage of the hens is typically considerably deteriorated, especially in the second part of the production period, because of wear and tear caused by the males (de Jong and Guemene, 2011; Jones and Prescott, 2000). A poor plumage increases the risk of skin damage, as the protective layer of feathers is diminished. The skin damage inflicted is typically found along the torso and thigh beneath the wings and on the back of the head and neck on the females (Fig. 5a–c) (Millman et al., 2000).

According to Millman et al. (2000), the broiler breeder industry associates the problem with breeder male aggression with one particular parent strain that came into use when the high breast-yielding lines were introduced. Subsequently, sexual aggression was reported in all commercial broiler breeder strains, suggesting that it was a by-product resulting from selection for sexual vigour or a result of genetic drift from the inbreeding of foundation lines (Millman et al., 2000). No studies on the extent of sexual aggression in broiler breeders have been published within the last decade, that is, up-to-date information is unavailable.

2.3 Mutilation and associated welfare problems

In order to limit the damage inflicted by bird-to-bird interactions, including within same sex, mutilation of specific body parts is practised in broiler breeders. Mutilation is the removal or damage to a part or parts of the body,

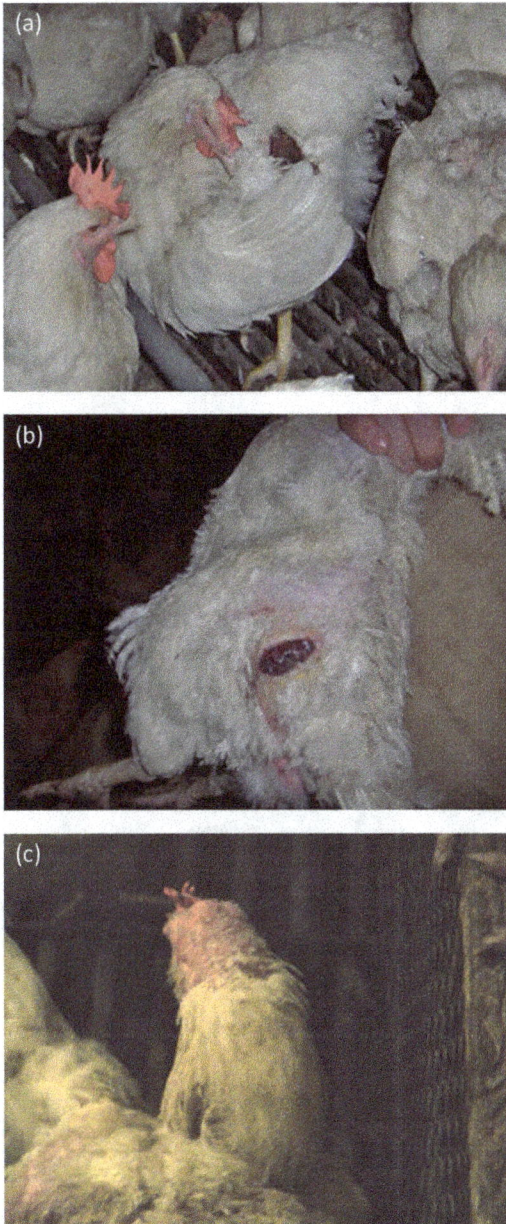

Figure 5 Examples of broiler breeder females having the typical injuries inflicted during mating by the males: Wounds along the torso (a) and thigh (b) that are likely to have been caused by the spur or toes of the males. The denuded skin on the back of the head (c) is likely to have been inflicted when the males during mating use the beak to grab the female by the neck or back of the head. ©Ingrid de Jong, Wageningen Livestock Research, The Netherlands.

not being the horny dead body tissue and feathers, due to either an operation or a trauma (van Niekerk and de Jong, 2007).

In broiler breeders, the body parts that are mutilated differ between countries (EFSA, 2010). Trimming of the beak is routinely practised in both sexes. Likewise, the removal of spurs and toe trimming are routinely practised in males, with the exception of some genotypes where the growth of the spurs is reduced, eliminating the need for spur trimming. Toe trimming is performed on the outermost joint of the toe pointing backwards, digit I, and occasionally also on digit II (Fig. 6). The final type of mutilation is dubbing of the male comb, but it is no longer commonly practised.

In a full-scale study on a commercial broiler breeder production farm, where the males were beak-intact and toe-intact, mortality and culls due to wounds of breeder females were found to be increased (de Jong et al., 2018). In addition, more damage to the skin was observed in the breeder females housed with beak-intact and toe-intact breeder males. The damage was most frequently located on the flank, thighs and back of the females, suggesting that the intact toes of the breeder males caused the damage.

Interestingly, research from the Netherlands indicates that beak-intact breeder females have reduced mortality, and in some EU countries (e.g. Poland and UK) beak-intact broiler breeders are housed without increased damage to the skin (van Emous and de Jong, 2013). A recent study showed that omitting beak trimming of the breeder males had no effect on feather damage of the

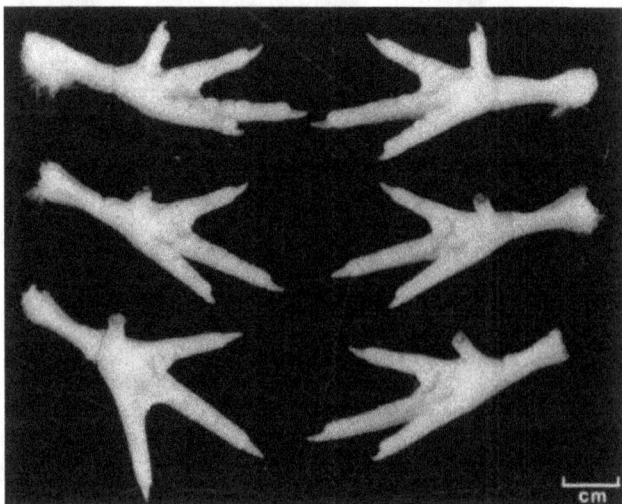

Figure 6 Left and right feet 24 hours after hatching of toe-intact chicks (top) and toe-trimmed chicks (digit I; middle and bottom). Toe trimming is performed immediately after hatching. ©(Gentle and Hunter, 1988).

breeder females but increased the feather damage of the breeder males (van Niekerk and de Jong, 2017).

Most research on the effect of the procedure of mutilation on bird welfare concerns beak trimming, whereas the other types of mutilations have received little attention. However, all mutilations are performed on living and thus innervated tissue, for which reason acute pain is very likely to be caused by all types of mutilations. The pain involved and the impact of the various mutilations may, however, vary, as the degree of innervation of the tissue mutilated may differ.

The beak of a domestic fowl is highly innervated and contains nociceptors, thermoreceptors and mechanoreceptors for sensation of pain, temperature and pressure/texture, respectively (Gentle, 1989). Beak trimming therefore results in pain and sensory loss (Gentle, 2011; Gentle et al., 1997). In addition, as the beak is a sensitive tool used during natural behaviour, such as for grasping food, preening and nest building, beak trimming is considered problematic, as it causes a reduction in the bird's ability to manipulate items. For instance, beak-trimmed laying hens have been observed to have higher infestations of ectoparasites than beak-intact individuals (Mullens et al., 2010; Chen et al., 2011; Vezzoli et al., 2015).

Innervation of toes in domestic fowl and the consequences in terms of pain involved when trimming toes are only addressed in one study. Gentle and Hunter (1988) investigated the damage inflicted due to toe trimming in broiler breeder males. The toe trimming procedure was done at hatch by cutting one-third of the backward-facing toes and did not involve cautery. The toes were found to be well-innervated, and some neural regeneration took place after toe trimming. Some regenerating nerves became trapped in scar tissue and formed small discrete neuromas, which persisted over the 60-day observation period. Gentle and Hunter (1988) concluded that toe trimming is likely to cause acute pain at the time of amputation and the immediate period following, but, compared to beak trimming, it is less likely to be followed by chronic pain.

Evidence of pain involved in toe trimming also comes from toe trimming of turkey toms. Fournier et al. (2015) showed that on day 1 after toe trimming, the toms spent more time resting, less time walking, standing and at the feeder than intact birds, indicating pain due to the toe trimming procedure. Some of these behavioural consequences persisted through till day 5. In addition to the pain, toe trimming may also cause difficulties in the performance of scratching behaviour and problems with balance. At 19 weeks of age, toe-trimmed turkey toms have been found to walk less than toe-intact birds, which may reflect difficulty of maintaining balance when lacking parts of the toes (Fournier et al., 2015). Whether toe trimming has similar effects on broiler breeder males remains to be investigated.

To summarise, beak trimming and toe trimming can lead to acute and chronic pain. Therefore, reducing the risk of damage to the broiler breeder females by mutilation of the males brings about yet another broiler breeder paradox - mutilation of the breeder males results in improved welfare of the females but reduced welfare of the males, whereas omitting mutilation results in reduced welfare of the females. With regard to beak trimming of broiler breeder females, it can be questioned whether there is a true need for this procedure, as the research, although sparse, seems to show that it makes no difference to the condition of the plumage and skin of the females, but the procedure itself has negative welfare consequences.

2.3.1 Methods of reducing aggression to eliminate the need for mutilation

Limited research aiming directly at finding alternatives to mutilation as a strategy of preventing damage due to aggression in broiler breeders has been conducted. However, there are studies that indirectly address this topic. For instance, provision of elevated resting places may improve the possibility of escaping aggression as discussed previously in this chapter. In this section, the few studies that have examined the possibilities of reducing the extent of sexual aggression are covered. These include investigations of the effect on occurrence of sexual aggression when providing panels, reducing the stocking density, using UV lighting or housing the sexes separately either part of the day or fulltime. Furthermore, the potential of breeding towards less-aggressive broiler breeders is briefly discussed.

2.3.1.1 The effect of panels

In the production system, broiler breeder males typically congregate on the littered area, while females appear fearful of the males and tend to avoid them by staying at the raised slats (Estevez, 1999; Millman et al., 2000). Leone and Estevez (2008) proposed that females in breeder flocks housed with increased possibilities of shelter in the littered area would have greater behavioural control over their environment, as the shelters would offer the opportunity to avoid aggressive interactions. As a result, male dispersal throughout the house would increase, and more females would therefore be attracted to the littered area, resulting in improved sexual interaction with fewer forced matings, less damage to females inflicted during mating and improved reproductive performance.

They investigated the hypothesis in an on-farm study where vertical cover panels were provided on the littered area centrally placed in the house during the production period (Leone and Estevez, 2008). The presence of cover panels was found to improve reproductive performance from 25 to 60 weeks of age. Egg production was increased by 2.1%, and hatchability and fertility improved, resulting in additional 4.5 chicks per female. The home range of the broiler breeder males housed with cover panels increased compared to males housed without panels. In addition, the broiler breeder males housed with cover panels were located more often on the slatted area than males housed without panels.

No observations of mating behaviour, damage inflicted on the broiler breeder females or the females' use of the littered area were recorded. Anecdotal results exist on these parameters from a demonstration (i.e. not an experiment in a controlled setting) where cover panels were introduced at 23 weeks of age in a commercial broiler breeder flock suffering from severe skin damage and high mortality of breeder females (Estevez, 1999). The introduction of cover panels was reported to have an immediate effect. Twenty-four hours after introducing the cover panels, the number of females in the scratching area increased by several hundred. The behaviour of the males was reported to be more relaxed, and mortality of the females was reduced.

Thus, providing cover panels seems to have the potential for reducing forced matings without compromising reproductive performance. Cover panels seem to improve the distribution within the house of both males and females whereby the mating opportunities of the males increases, and the damage and stress to the females caused by forced matings potentially may be decreased. As such, the provision of cover panels should be considered in the strategy for preventing damage inflicted on broiler breeder females during the production period.

2.3.1.2 The effect of stocking density

Kratzer and Craig (1980) found that layer breeder males showed more courtship behaviour at a reduced stocking density. They suggested that the lower stocking density facilitated an improvement of the females' recognition of male courtship behaviour. Following up on this hypothesis, de Jong et al. (2011) did an on-farm study where broiler breeders were housed at a standard stocking density of 8.8 birds/m^2 and a low stocking density of 5.2 birds/m^2. They found that a reduced stocking density resulted in more matings preceded by courtship behaviour (Fig. 7a), fewer forced matings (Fig. 7b) and more successful matings, that is, fewer hens struggled and escaped during mating (Fig. 7c). The effect was greatest if the stocking density was reduced during both the rearing period and the production period. The damage to the plumage and skin was less severe in both females and males housed at the reduced stocking density. Furthermore, more eggs were fertilised, more eggs hatched and a higher number of day-old chicks per hen were gained in the flocks housed at reduced stocking density.

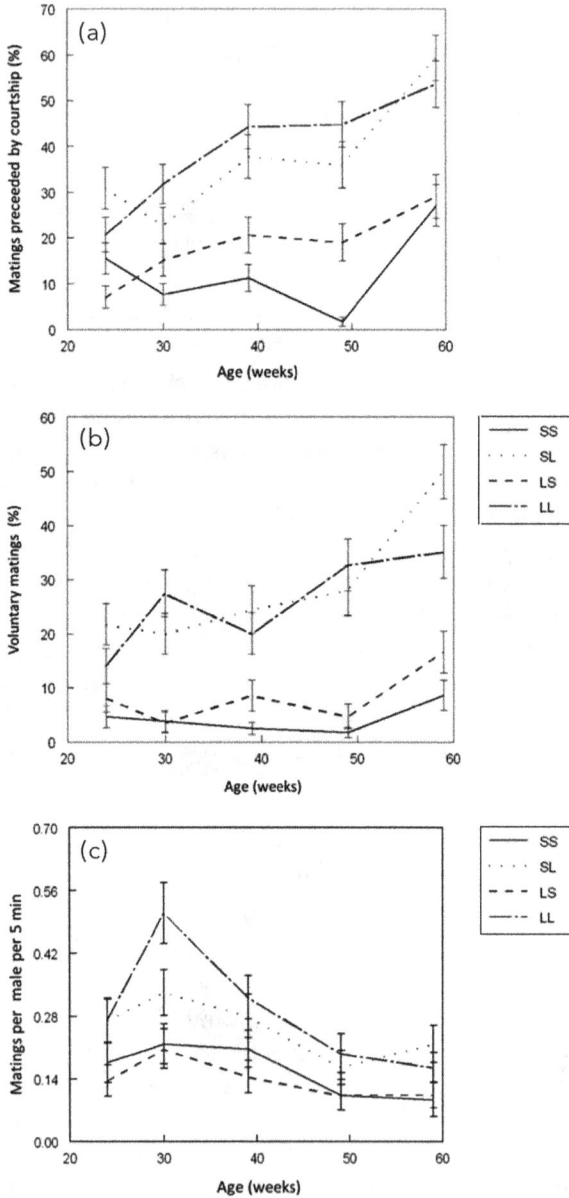

Figure 7 The effect of stocking density (SD) during both the rearing period and the production period on (a) percentage of matings preceded by courtship behaviour, (b) percentage of voluntary matings and (c) mean number of successful matings per breeder male per 5 min. SS = standard SD both during rearing and production, LL = low SD both during rearing and production, SL = standard SD during rearing, low SD during production, LS = low SD during rearing and standard SD during production. ©de Jong et al. (2011).

Thus, reducing stocking density increases the proportion of voluntary matings and thus reduces the risk of damage inflicted on the broiler breeder females during mating. In addition, reduced stocking density improves the condition of the plumage and skin of the females, which may further protect against damage inflicted during mating. Importantly, reducing the stocking density can be done without compromising reproductive performance. Reduced stocking density may be used as a management tool to lower the risk of damage inflicted during mating. However, economically, this solution appears costly.

2.3.1.3 The effect of ultraviolet light

Domestic fowls, including broiler breeders, are capable of seeing ultraviolet light with a wavelength between 320 and 400 mm, equalling UV_A light. From other bird species it is known that UV_A light plays a role for social communication and choice of partner through reflections of the plumage (see review in Jones et al., 2001). The light sources traditionally used in poultry houses do not include UV_A light.

In broiler breeders, the behaviour in a UV_A-enriched environment has been examined (Jones et al., 2001). The number of sexual attempts and locomotion was shown to increase in the environment enriched with UV_A light, and the females' choice of partner depended on the level of UV_A light. The conclusion was that UV_A light is involved in the communication of mating signals in broiler breeders and that UV_A light, therefore, potentially may influence the level of sexual aggression in broiler breeders.

2.3.1.4 Housing systems separating sexes

Using housing systems that separate the sexes either temporarily or permanently may be a solution to reduce/prevent the sexual aggression in broiler breeder production. In one system, the Quality Time Concept®, broiler breeder males and females are kept in large sex-mixed flocks, ensuring natural mating, but periodically the system separates the birds by taking advantage of the different types of feeders used for the two sexes. Another system is cages in which sexes are kept apart for lifetime, for which reason artificial insemination is necessary. Below, these two systems are described and discussed in terms of welfare implications to the broiler breeders.

The Quality Time Concept®: One alternative housing system for broiler breeders is the Quality Time Concept® (QTC; van Emous, 2010). The system was developed to adapt measures to improve fertility, based on variation in mating frequency over the day. In this housing system, broiler breeder males and females are separated for 5 hours a day during the light period, using separate feeding systems and a movable partition. When it is feeding time, the sexes move towards the feeders they are able to feed from, and while they are busy feeding, a partition is placed, separating the two sexes (Fig. 8).

Mating behaviour in the QTC compared to the standard housing system has been shown to be more voluntary, resulting in more successful matings (van Emous, 2010). In the QTC flocks, a better plumage of females aged 37 to 48 weeks was found, and no increase in aggressive behaviour among males occurred. In addition, increased mixing of males and females took place 2 hours before the dark period, indicating an improved sexual interaction. At this time period, mating activity normally peaks (Harris et al., 1980; Duncan et al., 1990; Bilcik and Estevez, 2005). In an initial trial, the fertility was reduced

Figure 8 The Quality Time Concept® (QTC) separating broiler breeder males and females daily for some hours during the light period. In the first version of the system (a) males are separated from females in three male enclosures at the long side of the house, whereas in the second improved version (b) all males are separated in the rear end of the house. ©Rick van Emous, Wageningen Livestock Research, The Netherlands.

by 1–2% in the QTC, but after improving the concept (Fig. 8b) the mean fertility rate increased (van Emous and Gunnink, 2011). Separating males from females was rather effective, as only 2.7% females and 5.2% males were found in the wrong compartment.

The cage system: In some parts of the world, broiler breeders are kept in single-sex cages during the production period. Artificial insemination is practised, and the need for despurring and toe trimming of breeder males to avoid injuries on females during mating is thereby eliminated. However, housing broiler breeders in cages may affect other aspects of welfare negatively. Fulfilling behavioural needs of domestic fowls is more difficult in cage systems (Lay et al., 2011), and it may not be acceptable to the consumers.

Artificial insemination may also stimulate selection of heavy males, which has negative consequences for male welfare (Laughlin, 2009; Brillard, 2001). In addition, artificial insemination involves regular handling of the birds. Often, the handling is rough/perceived as rough by the birds, which may have negative implications for bird welfare (Rushen et al., 1999; Laurence et al., 2014). The production of chicks may also be negatively affected by housing in cages. A study from 1983 showed a reduced fertility and hatchability of eggs from artificially inseminated-caged breeder females compared to naturally mated females (Petitte et al., 1983).

To summarise, QTC reduces forced matings and is likely to decrease the frequency of overmating. Consequently, the risk of inflicting damage on the breeder females is reduced. QTC may therefore be part of a solution to reduce damage to the broiler breeder females caused by sexual aggression. Housing broiler breeders in single-sex cages solves the issue with damage inflicted by broiler breeder males on the females during mating and eradicates the need for toe trimming and despurring, but it generates other welfare problems and may affect production of chicks negatively. In addition, semen collection from breeder males and artificial insemination of breeder females are cost-intensive solutions because of the very time-consuming nature.

2.3.1.5 Breeding for normal sexual behaviour

Studies have shown that the aggression displayed by broiler breeder males has a genetic component. Millman et al. (2000) compared layer and broiler breeder males and found that the latter displayed higher levels of both non-sexual and sexual aggression. Compared to layer breeder males, broiler breeder males chased and pecked females more (Fig. 9a). They also forced matings more than three times as often as did layer breeder males (Fig. 9b). Unsuccessful attempts of mating, often interrupted due to struggling females, leading to escape, were almost twice as common in broiler breeder males compared

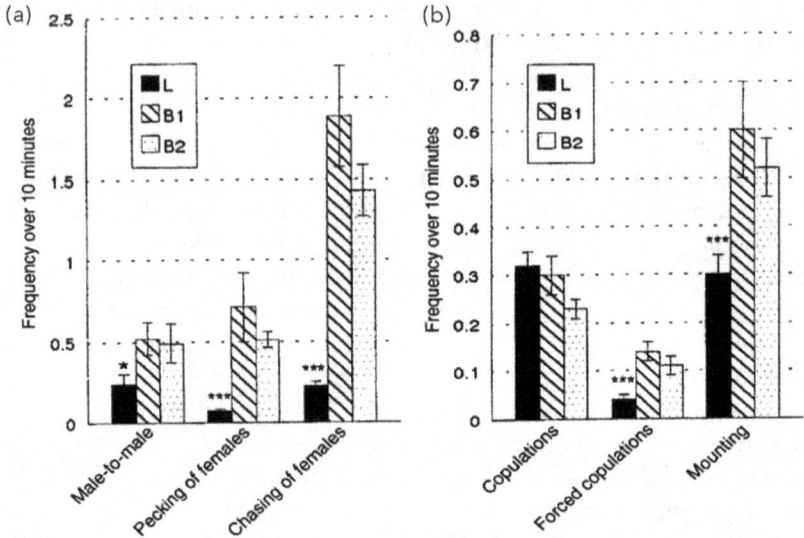

Figure 9 Differences between males from a strain of layer breeders (L: Isa Brown) and from two strains of broiler breeders (B1: Peterson, B2: Ross) in (a) aggressive behaviour and (b) mating behaviour (copulation = voluntary and forced matings; mounting = unsuccessful matings where the female escaped). ©Millman et al. (2000).

to layer breeder males (Fig. 9b). Since sexual aggression was not affected by feeding regimes (*ad libitum* fed vs. feed restricted), it was concluded that the differences between the broiler breeder strains and the layer breeder strain were associated with genetic factors.

The higher level of sexual aggression may also be related to the underdeveloped courtship behaviour displayed by broiler breeders. For instance, in the study by Millman et al. (2000), courtship behaviour in the two strains of broiler breeders was virtually absent before mating (Fig. 10). In the early stage of development of sexual behaviour, sexual aggression may also occur in both jungle fowl and layer breeders, but as males become sexually experienced, the aggressive elements disappear (Wood-Gush, 1958; Kruijt, 1964). However, this development in sexual behaviour is not observed in broiler breeder males, as sexual aggression persists even with attainment of considerable sexual experience (Millman et al., 2000).

Millman et al. (2000) suggested that it is possible that broiler breeder males are developmentally retarded and have become halted at an early stage of sexual development. Adding the fact that de Jong et al. (2011) found fewer forced matings when recognition of mating behaviour was improved, this indicates that performance of normal courtship behaviour may potentially lead to less damage inflicted on the breeder females during mating. Thus, breeding for normal sexual behaviour, including display of courtship behaviour

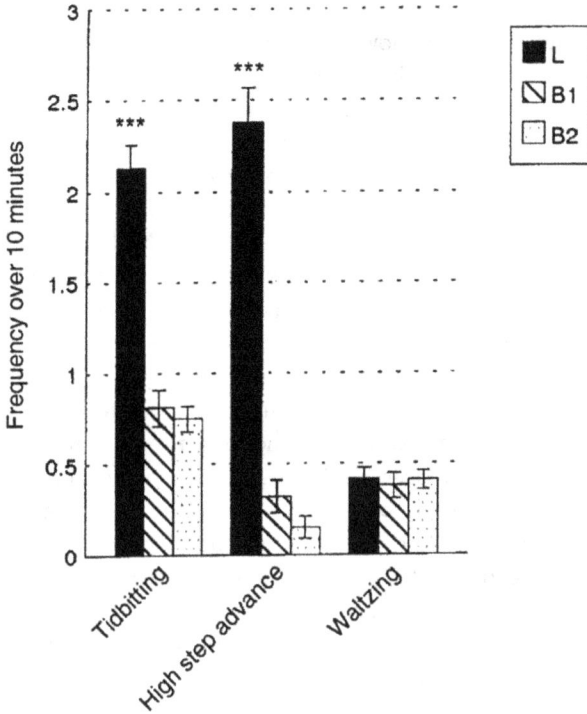

Figure 10 Differences between males from a strain of layer breeders (L: Isa Brown) and from two strains of broiler breeders (B1: Peterson, B2: Ross) in display of courtship behaviour. ©Millman et al. (2000).

and reduced sexual aggression, seems overly important and essential in the achievement of improved welfare of broiler breeder females.

3 Conclusion and future trends

It appears from the exposition of knowledge on welfare issues affecting broiler breeders that there are a number of significant challenges which need to be addressed in the near future. Among these are the severe feed restrictions and the associated welfare problems in terms of water restriction, physiological stress and development of abnormal behaviour, indicative of frustration and hunger. In addition to the unfulfilled behavioural need for ingesting feed, other behavioural needs, for example, for foraging and elevated resting, are also compromised. Furthermore, welfare challenges associated with the high level of aggression, particularly sexual aggression, exist, causing increased levels of fear, pain and mortality. Linked to this is the mutilation of broiler breeders, which is used as a preventive measure to reduce damage inflicted to the plumage and skin of flock mates, but the mutilation is a welfare problem in itself.

As mentioned previously, research into welfare issues affecting broiler breeders is rather limited. However, this may change to some extent, as political attention has been drawn to the topic, which may open up for more research. Examples include the attention that has been drawn to the mutilation procedure in the domestic fowl during recent years, mainly in Northern Europe and Northern America. With a few exceptions, focus has been on beak trimming in laying hens, but Denmark and the Netherlands have taken it a step further by including mutilation in broiler breeders. In the Netherlands, it has been politically decided that mutilation in broiler breeders should be phased out by 2021. Moreover, the authorities in some countries, for example, Denmark and the Netherlands, have initiated work aiming at finding solutions to reduce the welfare problems associated with feed restriction.

Consumer-driven forces may also result in more research into how to improve welfare of broiler breeders. During recent years, consumer awareness of animal welfare has increased worldwide, though particularly in Europe and North America. Concurrently, an increasing number of companies within the food industry have included or improved the policy on animal welfare in their corporate social responsibility (CSR). The consumer demand for organic or middle-segment broiler meat has resulted in alternative genotypes of broiler breeders with reduced growth rates that have become available in some parts of the world, although still at low numbers.

4 Acknowledgements

Part of this chapter originates from a report, 'Alternatives to mutilation of the outermost joint of the backward-facing toe in broiler breeder males', that was commissioned by the Ministry of Environment and Food of Denmark as part of the 'Contract between Aarhus University and Ministry of Environment and Food for the provision of research-based policy advice at Aarhus University, 2017-2020'. The report is available online at: https://pure.au.dk/portal/da/persons/anja-brinch-riber(b746c493-a840-48f9-af06-0e6ac28ae8ba)/publications/alternatives-to-mutilation-of-the-outermost-joint-of-the-backwardfacing-toe-in-broiler-breeder-males(71baf919-7137-4084-a8dc-866cd1265eb0).html.

5 Where to look for further information

Further information on welfare issues affecting broiler breeders may be found in the following review papers:

- D'Eath et al. (2009): 'Freedom from hunger' and preventing obesity: the animal welfare implications of reducing food quantity or quality.
- de Jong and Guemene (2011): Major welfare issues in broiler breeders.

- Mench (2002): Broiler breeders: feed restriction and welfare.
- Riber et al. (2017): Environmental enrichment for broiler breeders: An undeveloped field.

Also recommended is the report containing the EFSA Scientific Opinion on welfare aspects of the management and housing of the grandparent and parent stocks raised and kept for breeding purposes (EFSA, 2010). The report can be found freely available on the web.

6 References

Arrazola, A. 2018. The effect of alternative feeding strategies for broiler breeder pullets throughout the production cycle. PhD Thesis. The University of Guelph.

Aviagen. 2019a. *Ross308-308FF: Performance Objectives*. Available at: http://en.aviagen.com/assets/Tech_Center/Ross_Broiler/Ross308-308FF-BroilerPO2019-EN.pdf.

Aviagen. 2019b. *Ross Parent Stock Management Handbook*. Available at: http://en.aviagen.com/assets/Tech_Center/Ross_PS/RossPSHandBook2018.pdf.

Bilcik, B. and Estevez, I. 2005. Impact of male–male competition and morphological traits on mating strategies and reproductive success in broiler breeders. *Applied Animal Behaviour Science* 92(4), 307-23. doi:10.1016/j.applanim.2004.11.007.

Bowling, M., Forder, R., Hughes, R. J., Weaver, S. and Hynd, P. I. 2018. Effect of restricted feed intake in broiler breeder hens on their stress levels and the growth and immunology of their offspring. *Translational Animal Science* 2(3), 263-71. doi:10.1093/tas/txy064.

Brake, J., Keeley, T. P. and Jones, R. B. 1994. Effect of age and presence of perches during rearing on tonic immobility fear reactions of broiler breeder pullets. *Poultry Science* 73(9), 1470-4. doi:10.3382/ps.0731470.

Brillard, J. P. 2001. Future strategies for broiler breeders: an international perspective. *World's Poultry Science Journal* 57(3), 243-50. doi:10.1079/WPS20010017.

Bruggeman, V., Onagbesan, O., D'hondt, E., Buys, N., Safi, M., Vanmontfort, D., Berghman, L., Vandesande, F. and Decuypere, E. 1999. Effects of timing and duration of feed restriction during rearing on reproductive characteristics in broiler breeder females. *Poultry Science* 78(10), 1424-34. doi:10.1093/ps/78.10.1424.

Chen, B. L., Haith, K. L. and Mullens, B. A. 2011. Beak condition drives abundance and grooming-mediated competitive asymmetry in a poultry ectoparasite community. *Parasitology* 138(6), 748-57. doi:10.1017/S0031182011000229.

Cobb. 2018. *Cobb500 Broiler Performance and Nutrition Supplement*. Available at: https://www.cobb-vantress.com/assets/Cobb-Files/product-guides/bdc20a5443/70dec630-0abf-11e9-9c88-c51e407c53ab.pdf.

Cordiner, L. S. and Savory, C. J. 2001. Use of perches and nestboxes by laying hens in relation to social status, based on examination of consistency of ranking orders and frequency of interaction. *Applied Animal Behaviour Science* 71(4), 305-17. doi:10.1016/s0168-1591(00)00186-6.

Craig, W. 1918. Appetites and aversions as constituents of instincts. *Biological Bulletin* 34(2), 91-107.

D'Eath, R. B., Tolkamp, B. J., Kyriazakis, I. and Lawrence, A. B. 2009. 'Freedom from hunger' and preventing obesity: the animal welfare implications of reducing food quantity or quality. *Animal Behaviour* 77(2), 275–88. doi:10.1016/j.anbehav.2008.10.028.

Decuypere, E., Bruggeman, V., Everaert, N., Li, Y., Boonen, R., de Tavernier, J., Janssens, S. and Buys, N. 2010. The broiler breeder paradox: ethical, genetic and physiological perspectives, and suggestions for solutions. *British Poultry Science* 51(5), 569–79. doi:10.1080/00071668.2010.519121.

de Jong, I. C. and Guemene, D. 2011. Major welfare issues in broiler breeders. *World's Poultry Science Journal* 67(1), 73–82. doi:10.1017/S0043933911000067.

de Jong, I. C., van Voorst, S., Ehlhardt, D. A. and Blokhuis, H. J. 2002. Effects of restricted feeding on physiological stress parameters in growing broiler breeders. *British Poultry Science* 43(2), 157–68. doi:10.1080/00071660120121355.

de Jong, I. C., van Voorst, A. S. and Blokhuis, H. J. 2003. Parameters for quantification of hunger in broiler breeders. *Physiology and Behavior* 78(4–5), 773–83. doi:10.1016/s0031-9384(03)00058-1.

de Jong, I. C., Enting, H., van Voorst, A. and Blokhuis, H. J. 2005a. Do low-density diets improve broiler breeder welfare during rearing and laying? *Poultry Science* 84(2), 194–203. doi:10.1093/ps/84.2.194.

de Jong, I. C., Fillerup, M. and Blokhuis, H. J. 2005b. Effect of scattered feeding and feeding twice a day during rearing on indicators of hunger and frustration in broiler breeders. *Applied Animal Behaviour Science* 92(1–2), 61–76. doi:10.1016/j.applanim.2004.10.022.

de Jong, I. C., Wolthuis-Fillerup, M. and van Emous, R. A. 2009. Development of sexual behaviour in commercially-housed broiler breeders after mixing. *British Poultry Science* 50(2), 151–60. doi:10.1080/00071660802710124.

de Jong, I. C., Lourens, A., Gunning, H., Wokel, L. and van Emous, R. 2011. Effect of stocking density on (the development of) sexual behaviour and technical performance in broiler breeds. (In Dutch: Effect van bezettingsdichtheid op (de ontwikkeling van) het paargedrag en de technische resultaten bij vleeskuikenouderdieren). Wageningen UR Livestock Research.

de Jong, I., van Hattum, T. and Gunnink, H. 2018. Meer schade en uitval bij volledig onbehandelde hanen (In English: More damage and loss with completely untreated roosters). *Pluimveehouderij*, 8 November 2018, pp. 26–8.

Denbow, D. M. 1989. Peripheral and central control of food intake. *Poultry Science* 68(7), 938–47. doi:10.3382/ps.0680938.

Duncan, I. J. H., Hocking, P. M. and Seawright, E. 1990. Sexual behavior and fertility in broiler breeder domestic fowl. *Applied Animal Behaviour Science* 26(3), 201–13. doi:10.1016/0168-1591(90)90137-3.

EFSA. 2010. Scientific Opinion on welfare aspects of the management and housing of the grand-parent and parent stocks raised and kept for breeding purposes. *EFSA Journal* 8(7), 1667.

Elfick, D. 2010. 50 years of selection in the broiler breeder industry and beyond. In: Ravindran, V. (Ed.), *Proceedings of the NZ Poultry Industry Conference*. Monogastric Research Centre, Massey University, Palmerston North, New Zealand.

Estevez, I. 1999. Cover panels for chickens: a cheap tool that can help you. *Poultry Perspectives* 1, 4–6.

Fournier, J., Schwean-Lardner, K., Knezacek, T. D., Gomis, S. and Classen, H. L. 2015. The effect of toe trimming on behavior, mobility, toe length and other indicators of welfare in tom turkeys. *Poultry Science* 94(7), 1446–53. doi:10.3382/ps/pev112.

Gebhardt-Henrich, S. G. and Oester, H. 2014. Do broiler breeders prefer elevated sleeping sites? In: Estevez, I., Manteca, X., Marin, R. H. and Averós, X. (Eds), *Proceedings of the 48th Congres of the International Society for Applied Ethology*, Vittoria-Gasteiz, Spain.

Gebhardt-Henrich, S. G., Toscano, M. J. and Wurbel, H. 2017. Perch use by broiler breeders and its implication on health and production. *Poultry Science* 96(10), 3539–49. doi:10.3382/ps/pex189.

Gentle, M. J. 1989. Cutaneous sensory afferents recorded from the nervus intramandibularis of *Gallus gallus* var *domesticus*. *Journal of Comparative Physiology. A, Sensory, Neural, and Behavioral Physiology* 164(6), 763–74. doi:10.1007/bf00616748.

Gentle, M. J. 2011. Pain issues in poultry. *Applied Animal Behaviour Science* 135(3), 252–8. doi:10.1016/j.applanim.2011.10.023.

Gentle, M. J. and Hunter, L. H. 1988. Neural consequences of partial toe amputation in chickens. *Research in Veterinary Science* 45(3), 374–6. doi:10.1016/S0034-5288(18)30968-8.

Gentle, M. J., Hughes, B. O., Fox, A. and Waddington, D. 1997. Behavioural and anatomical consequences of two beak trimming methods in 1- and 10-d-old domestic chicks. *British Poultry Science* 38(5), 453–63. doi:10.1080/00071669708418022.

Girard, M. T. E., Zuidhof, M. J. and Bench, C. J. 2017a. Feeding, foraging, and feather pecking behaviours in precision-fed and skip-a-day-fed broiler breeder pullets. *Applied Animal Behaviour Science* 188, 42–9. doi:10.1016/j.applanim.2016.12.011.

Girard, T. E., Zuidhof, M. J. and Bench, C. J. 2017b. Aggression and social rank fluctuations in precision-fed and skip-a-day-fed broiler breeder pullets. *Applied Animal Behaviour Science* 187, 38–44. doi:10.1016/j.applanim.2016.12.005.

Gunnarsson, S., Yngvesson, J., Keeling, L. J. and Forkman, B. 2000. Rearing without early access to perches impairs the spatial skills of laying hens. *Applied Animal Behaviour Science* 67(3), 217–28. doi:10.1016/s0168-1591(99)00125-2.

Harris, G. C., Benson, J., Sellers, R. and Cole, T. A. 1980. The mating activity of broiler breeder cockerels. *Arkansas Farm Research* 29, 15.

Heck, A., Onagbesan, O., Tona, K., Metayer, S., Putterflam, J., Jego, Y., Trevidy, J. J., Decuypere, E., Williams, J., Picard, M. and Bruggeman, V. 2004. Effects of ad libitum feeding on performance of different strains of broiler breeders. *British Poultry Science* 45(5), 695–703. doi:10.1080/00071660400006537.

Hiemstra, S. J. and Napel, J. T. 2013. Study of the impact of genetic selection on the welfare of chickens bred and kept for meat production. Available at: https://ec.euro pa.eu/food/sites/food/files/animals/docs/aw:practice_farm_broilers_653020_final-report_en.pdf.

Hocking, P. M. 2004. Measuring and auditing the welfare of broiler breeders. In: Weeks, C. A. and Butterworth, A. (Eds), *Measuring and Auditing Broiler Welfare*. CABI Publishing, Wallingford, UK, pp. 19–35.

Hocking, P. M. 2006. High-fibre pelleted rations decrease water intake but do not improve physiological indexes of welfare in food-restricted female broiler breeders. *British Poultry Science* 47(1), 19–23. doi:10.1080/00071660500468041.

Hocking, P. M. and Jones, E. K. 2006. On-farm assessment of environmental enrichment for broiler breeders. *British Poultry Science* 47(4), 418–25. doi:10.1080/00071660600825074.

Hocking, P. M., Maxwell, M. H. and Mitchell, M. A. 1993. Welfare assessment of broiler breeder and layer females subjected to food restriction and limited access to water during rearing. *British Poultry Science* 34(3), 443–58. doi:10.1080/00071669308417600.

Hocking, P. M., Maxwell, M. H. and Mitchell, M. A. 1996. Relationships between the degree of food restriction and welfare indices in broiler breeder females. *British Poultry Science* 37(2), 263–78. doi:10.1080/00071669608417858.

Hocking, P. M., Maxwell, M. H., Robertson, G. W. and Mitchell, M. A. 2001. Welfare assessment of modified rearing programmes for broiler breeders. *British Poultry Science* 42(4), 424–32. doi:10.1080/00071660120070677.

Hocking, P. M., Maxwell, M. H., Robertson, G. W. and Mitchell, M. A. 2002. Welfare assessment of broiler breeders that are food restricted after peak rate of lay. *British Poultry Science* 43(1), 5–15. doi:10.1080/00071660120109818.

Hocking, P. M., Channing, C. E., Robertson, G. W., Edmond, A. and Jones, R. B. 2004a. Between breed genetic variation for welfare-related behavioural traits in domestic fowl. *Applied Animal Behaviour Science* 89(1-2), 85–105. doi:10.1016/j.applanim.2004.03.014.

Hocking, P. M., Zaczek, V., Jones, E. K. M. and Macleod, M. G. 2004b. Different concentrations and sources of dietary fibre may improve the welfare of female broiler breeders. *British Poultry Science* 45(1), 9–19. doi:10.1080/0007166041000 1668806.

Hocking, P. M., Jones, E. K. M. and Picard, M. 2005. Assessing the welfare consequences of providing litter for feed-restricted broiler breeders. *British Poultry Science* 46(5), 545–52. doi:10.1080/00071660500254813.

Horne, P. L. M. V. 2018. *Competitiveness of the EU Poultry Meat Sector, Base Year 2017.* International Comparison of Production Costs, Wageningen, the Netherlands.

Jensen, P. and Toates, F. M. 1993. Who needs 'behavioural needs'? Motivational aspects of the needs of animals. *Applied Animal Behaviour Science* 37(2), 161–81. doi:10.1016/0168-1591(93)90108-2.

Jones, E. K. M. and Prescott, N. B. 2000. Visual cues used in the choice of mate by fowl and their potential importance for the breeder industry. *World's Poultry Science Journal* 56(2), 127–38. doi:10.1079/WPS20000010.

Jones, E. K. M., Prescott, N. B., Cook, P., White, R. P. and Wathes, C. M. 2001. Ultraviolet light and mating behaviour in domestic broiler breeders. *British Poultry Science* 42(1), 23–32. doi:10.1080/713655008.

Jones, E. K., Zaczek, V., Macleod, M. and Hocking, P. M. 2004. Genotype, dietary manipulation and food allocation affect indices of welfare in broiler breeders. *British Poultry Science* 45(6), 725–37. doi:10.1080/00071660400014226.

Kostal, L., Savory, C. J. and Hughes, B. O. 1992. Diurnal and individual variation in behaviour of restricted-fed broiler breeders. *Applied Animal Behaviour Science* 32(4), 361–74. doi:10.1016/S0168-1591(05)80028-0.

Kratzer, D. D. and Craig, J. V. 1980. Mating behaviour of cockerels: effects of social status, group size and group densitiy. *Applied Animal Ethology* 6(1), 49–62. doi:10.1016/0304-3762(80)90093-0.

Kruijt, J. P. 1964. Ontogeny of social behaviour in Burmase red junglefowl (Gallus gallus spadiceus) bonnaterre. *Behaviour* Suppl. XII, 1–201.

Laughlin, K. F. 2009. Breeder management: how did we get here? In: Hocking, P. (Ed.), *Biology of Breeding Poultry. Conference: 29th Poultry Science Symposium*, Edinburgh, Scotland, 23–25 July 2007.

Laurence, A., Lumineau, S., Calandreau, L., Arnould, C., Leterrier, C., Boissy, A. and Houdelier, C. 2014. Short- and long-term effects of unpredictable repeated negative stimuli on Japanese quail's fear of humans. *PLoS ONE* 9(3), e93259. doi:10.1371/journal.pone.0093259.

Lay, D. C., Fulton, R. M., Hester, P. Y., Karcher, D. M., Kjaer, J. B., Mench, J. A., Mullens, B. A., Newberry, R. C., Nicol, C. J., O'Sullivan, N. P. and Porter, R. E. 2011. Hen welfare in different housing systems. *Poultry Science* 90(1), 278–94. doi:10.3382/ps.2010-00962.

Leone, E. H. and Estevez, I. 2008. Economic and welfare benefits of environmental enrichment for broiler breeders. *Poultry Science* 87(1), 14–21. doi:10.3382/ps.2007-00154.

Lindholm, C., Johansson, A., Middelkoop, A., Lees, J. J., Yngwe, N., Berndtson, E., Cooper, G. and Altimiras, J. 2018. The quest for welfare-friendly feeding of broiler breeders: effects of daily vs. 5:2 feed restriction schedules. *Poultry Science* 97(2), 368–77. doi:10.3382/ps/pex326.

Mason, G. J. 1991. Stereotypies: a critical review. *Animal Behaviour* 41(6), 1015–37. doi:10.1016/S0003-3472(05)80640-2.

Mench, J. A. 1988. The development of aggressive behavior in male broiler chicks: a comparison with laying-type males and the effects of feed restriction. *Applied Animal Behaviour Science* 21, 233–42.

Mench, J. A. 2002. Broiler breeders: feed restriction and welfare. *World's Poultry Science Journal* 58(1), 23–9. doi:10.1079/WPS20020004.

Merlet, F., Puterflam, J., Faure, J. M., Hocking, P. M., Magnusson, M. S. and Picard, M. 2005. Detection and comparison of time patterns of behaviours of two broiler breeder genotypes fed ad libitum and two levels of feed restriction. *Applied Animal Behaviour Science* 94(3–4), 255–71. doi:10.1016/j.applanim.2005.02.014.

Millman, S. T. and Duncan, I. J. H. 2000. Effect of male-to-male aggressiveness and feed-restriction during rearing on sexual behaviour and aggressiveness towards females by male domestic fowl. *Applied Animal Behaviour Science* 70(1), 63–82. doi:10.1016/s0168-1591(00)00141-6.

Millman, S. T., Duncan, I. J. H. and Widowski, T. M. 2000. Male broiler breeder fowl display high levels of aggression toward females. *Poultry Science* 79(9), 1233–41. doi:10.1093/ps/79.9.1233.

Moradi, S., Zaghari, M., Shivazad, M., Osfoori, R. and Mardi, M. 2013. Response of female broiler breeders to qualitative feed restriction with inclusion of soluble and insoluble fiber sources. *Journal of Applied Poultry Research* 22(3), 370–81. doi:10.3382/japr.2012-00504.

Morrissey, K. L., Widowski, T., Leeson, S., Sandilands, V., Arnone, A. and Torrey, S. 2014. The effect of dietary alterations during rearing on feather condition in broiler breeder females. *Poultry Science* 93(7), 1636–43. doi:10.3382/ps.2013-03822.

Mullens, B. A., Chen, B. L. and Owen, J. P. 2010. Beak condition and cage density determine abundance and spatial distribution of northern fowl mites, Ornithonyssus sylviarum, and chicken body lice, *Menacanthus stramineus*, on caged laying hens. *Poultry Science* 89(12), 2565-72. doi:10.3382/ps.2010-00955.

Nielsen, B. L., Thodberg, K., Malmkvist, J. and Steenfeldt, S. 2011. Proportion of insoluble fibre in the diet affects behaviour and hunger in broiler breeders growing at similar rates. *Animal: an International Journal of Animal Bioscience* 5(8), 1247-58. doi:10.1017/S1751731111000218.

Norman, K. I., Weeks, C. A., Pettersson, I. C. and Nicol, C. J. 2018. The effect of experience of ramps at rear on the subsequent ability of layer pullets to negotiate a ramp transition. *Applied Animal Behaviour Science* 208, 92-9. doi:10.1016/j.applanim.2018.08.007.

OECD/FAO. 2018. World meat projections. *OECDFAO Agricultural Outlook 2018-2027.*

Ordas, B., Vahedi, S., Seidavi, A., Rahati, M., Laudadio, V. and Tufarelli, V. 2015. Effect of testosterone administration and spiking on reproductive success of broiler breeder flocks. *Reproduction in Domestic Animals = Zuchthygiene* 50(5), 820-5. doi:10.1111/rda.12595.

Petitte, J. N., Hawes, R. O. and Gerry, R. W. 1983. The influence of cage versus floor pen management of broiler breeder hens on subsequent performance of cage reared broilers. *Poultry Science* 62(7), 1241-6. doi:10.3382/ps.0621241.

Pollock, D. L. 1999. A geneticist's perspective from within a broiler primary breeder company. *Poultry Science* 78(3), 414-8. doi:10.1093/ps/78.3.414.

Puterflam, J., Merlet, F., Faure, J. M., Hocking, P. M. and Picard, M. 2006. Effects of genotype and feed restriction on the time-budgets of broiler breeders at different ages. *Applied Animal Behaviour Science* 98(1-2), 100-13. doi:10.1016/j.applanim.2005.08.013.

Renema, R. A. and Robinson, F. E. 2004. Defining normal: comparison of feed restriction and full feeding of female broiler breeders. *World's Poultry Science Journal* 60(4), 508-22. doi:10.1079/WPS200434.

Renema, R. A., Rustad, M. E. and Robinson, F. E. 2007. Implications of changes to commercial broiler and broiler breeder body weight targets over the past 30 years. *World's Poultry Science Journal* 63(3), 457-72. doi:10.1017/S0043933907001572.

Riber, A. B., de Jong, I. C., vad de Weerd, H. A. and Steenfeldt, S. 2017. Environmental enrichment for broiler breeders: an undeveloped field. *Frontiers in Veterinary Science* 4, 86. doi:10.3389/fvets.2017.00086.

Rushen, J., Taylor, A. A. and de Passillé, A. M. 1999. Domestic animals' fear of humans and its effect on their welfare. *Applied Animal Behaviour Science* 65(3), 285-303. doi:10.1016/S0168-1591(99)00089-1.

Sandilands, V., Tolkamp, B. J., Savory, C. J. and Kyriazakis, I. 2006. Behaviour and welfare of broiler breeders fed qualitatively restricted diets during rearing: are there viable alternatives to quantitative restriction? *Applied Animal Behaviour Science* 96, 53-67.

Savory, C. J. and Maros, K. 1993. Influence of degree of food restriction, age and time of day on behavior of broiler breeder chickens. *Behavioural Processes* 29(3), 179-89. doi:10.1016/0376-6357(93)90122-8.

Savory, C. J., Hocking, P. M., Mann, J. S. and Maxwell, M. H. 1996. Is broiler breeder welfare improved by using qualitative rather than quantative food restriction to limit growth rate? *Animal Welfare* 5, 105-27.

Shea, M. M., Mench, J. A. and Thomas, O. P. 1990. The effect of dietary tryptophan on aggressive behavior in developing and mature broiler breeder males. *Poultry Science* 69(10), 1664-9. doi:10.3382/ps.0691664.

Siegel, P. B. and Wolford, J. H. 2003. A review of some results of selection for juvenile body weight in chickens. *Journal of Poultry Science* 40(2), 81-91. doi:10.2141/jpsa.40.81.

Steenfeldt, S. and Nielsen, B. L. 2012. Effects of high fibre diets on gut fill, behaviour and productivity in broiler breeders. *2nd JRS Poultry Seminar*, Hannover, Germany.

Stratmann, A., Frohlich, E. K. F., Harlander-Matauschek, A., Schrader, L., Toscano, M. J., Wurbel, H. and Gebhardt-Henrich, S. G. 2015. Soft perches in an aviary system reduce incidence of keel bone damage in laying hens. *PLoS ONE* 10(3), e0122568. doi:10.1371/journal.pone.0122568.

van Emous, R. A. 2010. Quality Time®: an innovative housing concept for broiler breeders. *The 2nd International Nutrition Symposium*, Wageningen, the Netherlands.

van Emous, R. A. and de Jong, I. C. 2013. Promising management measures to solve welfare problems in Broiler Breeders. *2nd International Poultry Meat Congress*, Antalya, Turkey.

van Emous, R. and Gunnink, H. 2011. *Innovative Broiler Breeder Housing System: 'Quality Time' House* [In Dutch: Innovatieve huisvesting voor vleeskuikenouderdieren: "Quality Time" stal). Wageningen UR Livestock Research, Wageningen, The Netherlands.

van Emous, R. A., Kwakkel, R., van Krimpen, M. and Hendriks, W. 2014. Effects of growth pattern and dietary protein level during rearing on feed intake, eating time, eating rate, behavior, plasma corticosterone concentration, and feather cover in broiler breeder females during the rearing and laying period. *Applied Animal Behaviour Science* 150, 44-54. doi:10.1016/j.applanim.2013.10.005.

van Niekerk, T. F. and de Jong, I. 2007. Mutilations in poultry in European poultry production systems. *Lohmann Information* 42, 35-46.

van Niekerk, T. C. G. M. and de Jong, I. 2017. *Stand van zaken achterwege laten van ingrepen bij pluimvee*. Wageningen Livestock Research, Wageningen, The Netherlands.

Vezzoli, G., Mullens, B. A. and Mench, J. A. 2015. Relationships between beak condition, preening behavior and ectoparasite infestation levels in laying hens. *Poultry Science* 94(9), 1997-2007. doi:10.3382/ps/pev171.

William, J. 2012. Ovarian dysfunction in broiler breeder hens. *Medycyna Weterynaryjna* 68, 131-8.

Wood-Gush, D. G. M. 1958. The effect of experience on the mating behaviour of the domestic cock. *Animal Behaviour* 6(1-2), 68-71. doi:10.1016/0003-3472(58)90010-1.

Zuidhof, M. J. 2018. Lifetime productivity of conventionally and precision-fed broiler breeders. *Poultry Science* 97(11), 3921-37. doi:10.3382/ps/pey252.

Zuidhof, M. J., Robinson, F. E., Feddes, J. J. R., Hardin, R. T., Wilson, J. L., McKay, R. I. and Newcombe, M. 1995. The effects of nutrient dilution on the well-being and performance of female broiler breeders. *Poultry Science* 74(3), 441-56. doi:10.3382/ps.0740441.

Zuidhof, M. J., Fedorak, M. V., Ouellette, C. A. and Wenger, I. I. 2017. Precision feeding: innovative management of broiler breeder feed intake and flock uniformity. *Poultry Science* 96(7), 2254-63. doi:10.3382/ps/pex013.

Chapter 5

Poultry welfare monitoring: group-level technologies

Marian Stamp Dawkins and Elizabeth Rowe, University of Oxford, UK

1 Introduction

In one sense, the health and welfare of commercial poultry is already highly automated. Modern broiler houses, for example, have sophisticated climate control systems that keep birds within the temperature and humidity guidelines recommended by the breeding companies (Yahav et al., 2005; Jones et al., 2005), delivery of food and water is automated and alarms are fitted to ensure a constant supply and light is regulated for both intensity and photoperiod (Corkery et al., 2013). There are environmental sensors to measure temperature (D'Alfonso et al., 1996; Carvalho et al., 2013; Coelho et al., 2016; Curi et al., 2017), ambient dust (Zhao et al., 2009), relative humidity (Carvalho et al., 2013; Coelho et al., 2016; Curi et al., 2017), vibration (Chen et al., 2010), ammonia concentration (Ji et al., 2016) and carbon dioxide concentration (Carvalho et al., 2013; Ji et al., 2016). In other words, the birds' basic needs for food, water and physical comfort can be catered for automatically (Fournel et al., 2017), thus fulfilling at least some of the requirements for good welfare as set out in the Five Freedoms (Brambell, 1965; Webster, 2001), Welfare Quality® (2009), World Organization for Animal Health General Principles (Fraser et al., 2013) and other welfare systems.

But, in another sense, welfare is very far from being automated for any poultry. Welfare is not just defined by the environment provided for an animal (inputs) but what the welfare outcomes actually are (Main et al., 2014).

http://dx.doi.org/10.19103/AS.2020.0078.07

Measuring health outcomes such as hockburn and pododermatitis in broilers at the slaughterhouse (Vanderhasselt et al., 2013) is a step in the right direction, but there are many aspects of welfare that simply cannot be measured on dead birds, such as whether they are able to walk without difficulty (Kestin et al., 1992). Currently, the welfare of living birds is generally assessed by human auditors visiting poultry houses and making observations (Ekstrand et al., 1998; de Jong et al., 2012) but this is very labour intensive and gives only a 'snapshot' of the state of the flock on the day of the visit (Wathes et al., 2008; Winckler, 2019). It can also risk spreading disease if the same team visits different sites. What would improve poultry welfare generally would be automated ways of measuring welfare outcomes that can be routinely applied to all flocks. Automated monitoring systems allow for non-intrusive assessment of welfare, where information can be collected without the stress of disturbing or handling animals (Wathes et al., 2008). Continuous automated monitoring can also provide a more complete picture of the overall welfare state of animals, and alerting farmers to problems as they arise in real time allows for fast and targeted interventions which will benefit the current flock, compared to traditional welfare assessments that occur at the end of the production cycle (Winckler, 2019; Buscher, 2019). This would benefit producers, consumers and the birds themselves.

In addition to these advantages, automated measures of welfare are often seen as more 'objective' than the 'subjective' assessment of a human (Banhazi et al., 2012). While this is potentially a benefit, it is important not to confuse the objectivity of the measurement with how these measurements are to be interpreted in welfare terms, which often contains a subjective or at least a contestable element of what is desirable. Certainly machines can more accurately and more continuously record what animals do than a person observing with the naked eye and writing down what they see. In the limited sense of being more accurate, then, automated measurements are indeed more objective. But the problem of interpreting the data that has been collected and deciding whether or not that is a good measure of welfare remains and is still a matter of controversy. Machine learning for welfare outcomes still needs to have specified what a 'good' or desirable welfare outcome is (Morota et al., 2018). In exactly the same way that the use of machine learning in human medicine needs to be related to desirable outcomes such as longer life, ability to walk or the satisfaction of patients, so too automated assessment of animal welfare still has to have an end-point – what is seen as good or desirable. Part of the reason that automated measures seem objective is that they force us humans to be clearer than we otherwise might be about what our welfare goals are, because we then have to instruct machines what to look for.

So what exactly do we mean by welfare, and, more importantly, in the context of this chapter, can machines measure it? Although there is no universal agreement about how to define animal welfare (Mason and Mendl,

1993; Fraser, 2008; Mellor and Beausoleil, 2015), it is possible to discern some common threads amongst the various definitions that are in use. An emphasis on good physical health, for example, is universally accepted as the foundation of good animal welfare. For example, the Five Freedoms (Brambell, 1965; Webster, 2001), a widely used system for assessing welfare around the world, lists freedom from disease and injury as one of the key indicators of welfare. This is emphasized just as strongly in more recent versions such as the Ten General Principles (OIE, 2016; Fraser et al., 2013), the Five Provisions or Domains (Mellor, 2016) and the Four Principles put forward by the European Welfare Quality assessment (Welfare Quality, 2009). Furthermore, many of the other criteria used by these and other welfare schemes are specifically aimed at maintaining good health - such as making sure that animals have adequate food and water and are kept in safe comfortable environments in which they are not injured. Many of the most pressing animal welfare issues such as feather-pecking in laying hens (Gunnarsson et al., 1999) are seen as serious precisely because of the injuries that result. Conversely, finding ways of preventing disease and stopping hens from injuring themselves by breaking their bones would be widely agreed as improving their welfare. Death, injury and disease are clear welfare outcomes with results that can be measured objectively and are therefore particularly suited to automation.

Of course, there is more to good welfare than simply the absence of disease and injury, as emphasized even more strongly in ideas such as Quality of Life (Bono and Mori, 2005; Broom, 2007) and lives worth living (Wathes, 2010; Mellor, 2016), but, as with humans, these more complex ideas are difficult to define and even more difficult to capture in algorithms (Scott et al., 2007; Taylor and Mills, 2007). This chapter will therefore be mainly based on measures that are the easiest to automate, such as lameness; mortality; disease; leg, foot and skin damage; and vocalizations. Then, as machine algorithms are improving all the time (Nasirahmadi et al., 2017; Liakos et al., 2018), this chapter will additionally look at the steps that have been taken to automate the measurement of other, more complex measures of welfare such as behaviour. The chapter will use examples largely, although not exclusively, from applications to broiler (meat) chickens because this is where the group-level technology has been most developed.

2 Types of automated assessment

2.1 Introduction: group versus individual

Unlike larger animals such as dairy cows or sows where each animal can be monitored and treated individually (Caja et al., 2016; Buscher, 2019), commercial poultry are kept in groups of many thousands and treated as a group (Winckler, 2019). Feed, vaccinations, medication, drinker height, lighting and other factors are not administered to a single individual but set or adjusted to the needs of the whole flock. Even when a particular bird is observed (e.g. during gait scoring; Kestin et al., 1992), this is only a part of a sample to give an estimate of a whole flock score. Individual chickens can be marked or tagged as part of research projects and yield valuable results about individual behaviour (Campbell et al., 2017; Stadig et al., 2018). For routine commercial farming, however, tagging all birds in a flock of either broilers or layers would present major logistical problems in applying and retrieving the tags, while computer identification and tracking of thousands of birds would be both computationally difficult and expensive. Even if such logistical problems could be overcome, it would currently be impossible to pick up and medicate a single bird among thousands. The large groups in which commercial broilers are kept have therefore led to group-level measures of welfare and to whole group approaches to automating welfare monitoring (Ben Sassi et al., 2016; Berckmans, 2017; Buscher, 2019). It is these group-level measures that are discussed here. In discussing the sensors used for such welfare monitoring (most commonly visual, sound or thermal imaging), we will ask two questions: what are the sensors detecting and how valid is the interpretation of this data in welfare terms? In general, finding sensors to collect data is the easy part. Turning the vast quantities of data that the sensors collect into output that is meaningful in welfare terms is much more difficult.

2.2 Visual information from CCTV or video

There are several different approaches to the processing of information coming from cameras. One is to use digital image processing technology to distinguish individual chickens from their background and then to analyse different postures of the birds to identify sick or healthy individuals (Zhuang et al., 2018); to identify particular behaviours (Leroy et al., 2006) such as wing spreading, scratching and preening (Pereira et al., 2013); to track their location in preference tests (Kashiha et al., 2015); and even to detect abnormalities of body oscillation, step frequency and step length in lame broilers (Aydin, 2017a). An alternative approach is not to recognize individual birds as such but to simply see a flock of birds as something that makes patterns of light against a darker background and then to analyse the patterns as the rate of change of brightness (or 'optical flow') in different parts of a moving visual image (Fleet and Weiss, 2005; Fig. 1). This gives a flock-level, rather than an individual bird, measure but has the great advantage that whole frames containing tens or hundreds of individual birds can be assessed together, and the statistical properties of these flow patterns can be derived automatically and continuously (Sonka et al., 1999) using algorithms that are simple enough to deliver results continuously in real time. Clustering of birds can be used to detect that they

| (a) Image frame at time t | (b) Image frame at time $t + 1$ |

Figure 1 'Optical flow' is the rate of change of image brightness in different parts of a moving visual image, measured by comparing patterns of light and dark in successive video frames over time (Fleet and Weiss, 2005). These changes are combined to give an estimate of local velocity vectors. If all white chickens on the darker background remain stationary from one image to the next, there will no change in the pattern of brightness and no 'flow'. But if some of the chickens move between frames, some white areas will become dark and *vice versa*, and this is registered as a net 'flow' (Dawkins et al., 2009).

are too cold (Pereira et al., 2012). Yet other approaches involve a combination of individual and group measures so that 'events' such as feather pulling can be recognized but not which individual bird was responsible (Lee et al., 2011).

Using one or more of these techniques, it is now possible to automate the assessment of basic health measures and to validate the results against welfare outcomes assessed in more conventional ways. For example, the eYenamic system takes camera image data every minute, calculates activity and occupation indices for different areas of the screen and is able to link deviations in these indices to welfare assessment of hockburn, pododermaitis and gait score as carried by human observers (Aydin et al., 2015; De Montis et al., 2013; Norton et al., 2016; Aydin, 2017b; Fernandez et al., 2018).

Also using commercial flocks of broiler chickens, Dawkins et al. (2009, 2012) showed that flock movements, as measured by the statistical properties of optical flow patterns, were correlated with key physical health outcomes. Flocks that had lower mean levels of movement and higher kurtosis (a measure of variability) were scored with higher levels of hockburn and foot pad dermatitis at the end of life. Furthermore, these differences were already apparent in young chicks of less than 4 days old (Roberts et al., 2012; Dawkins et al., 2017) before there were any external signs of leg or foot damage. Optical flow patterns were also shown to differ between flocks subsequently testing positive or negative for *Campylobacter*, one of the major bacterial sources of food poisoning in humans, when chicks were less than a week old (Colles et al., 2016). *Campylobacter*-positive flocks showed lower mean and higher kurtosis of optical flow than flocks that remained free of infection, in line with the previous result that good welfare, including freedom from disease, is associated with more movement (higher mean) and more uniform movement (lower kurtosis). Since infection with *Campylobacter* is not detectable by normal culture methods until birds are around 21 days old, optical flow patterns were not only correlated with bird health but were also predicting future health status.

A similar ability of optical flow patterns to predict problems before they have detectable clinical effects is suggested by a study by Lee et al. (2011) on feather damage in laying hens. Optical flow patterns of the behaviour of 18-week-old birds were analysed from videos, this time using a hidden Markov model to quantify the number of anomalies or unusual events. The results correlated well with those recorded by a human observer and were predictive of the amount of feather damage when the same flocks were 38 weeks old. Feather damage is often a precursor for outbreaks of feather-pecking (Drake et al., 2010), so these preliminary results suggest that there is a potential for using automated monitoring for giving the warning of flocks at risk of welfare problems.

2.3 Sound

The technology to record and analyse sound is well developed, and sound can be used in welfare assessment in a variety of ways. The most obvious is to record sounds made by the birds themselves (Zimmerman et al., 2000). For example, very young chicks are unable to regulate their own body temperature and give high-pitched or distress calls when they are cold (Wood-Gush, 1971). By putting individual chicks into a climate-controlled chamber and altering the temperature, Moura et al. (2008) showed that both the amplitude and the frequency of these calls increased when the temperature dropped. In farm conditions, heat stress leads to distinctive calls by both broilers (Pereira et al., 2014) and layers (Lee et al., 2015). Du et al. (2018) monitored the number of vocalisations/bird both during the day and at night and argued that anomalies in this number of calls given by each bird, particularly at night, could be used for welfare assessment.

Another approach is to use non-communicatory sounds such as the noise made when birds peck food. In one study, a microphone positioned on the side of a feeder was used to estimate when and how much birds were eating (Aydin and Berckmans, 2016).

2.4 Temperature

Infra-red thermography is a non-invasive technique for measuring body surface temperature which has been related to stress, emotional arousal and leg pathologies in many bird species, including poultry (McCafferty, 2013). Changes in the surface temperature of the areas of bare skin can be symptomatic of short-term stress. For example, Edgar et al. (2013) showed that handling individual hens lead to an initial decrease of as much as 2°C in the surface head, eye and comb temperatures followed by an increase. Moe et al. (2017) found that during a 10-minute restraint, footpad temperatures dropped by nearly 0.5°C, while head region temperatures (e.g. nostrils, comb, wattle, eye) rose by 0.76°C. Both of these studies suggest that surface temperature changes can be used as measures of welfare but both were carried out in individual birds. Successfully translating these individual measurements into flock-level measures of 'stress' and distinguishing these from changes in the ambient temperature have yet to be achieved. Thermal imaging can also be used to assess feather cover as feathers insulate a bird and prevent heat loss. Cook et al. (2006) showed that the percentage of a thermal image that was within the temperature range of 17-24 C was negatively associated with feather cover, while the percentage within the range of 28-31 C was associated with a lack of feather cover or bare skin. Thermographic images have also been used to register multiple occupancies by laying hens, and the analysis of thermal images is now able count the number of birds in a given nest (Zaninelli et al., 2016, 2017, 2018).

As well as being used to assess the physical state of the birds themselves, thermal imaging cameras can also identify, with high precision, areas in a chicken house with different radiant temperatures and so alert the farmer to particular problems (Coelho et al., 2016).

2.5 Conclusions

It is clear that there are now, at least in the development stage, an increasing number of ways in which the welfare of poultry can be monitored automatically. Different sensors give different sorts of information, but for this to be used to maximum effect, it will be necessary to integrate the different sorts of information that are now available and to display them in a form that is accessible to farmers and easy for them to use (Van Hertem et al., 2017; Fournel et al., 2017). The full potential of many of these methods to give an automated picture of welfare has yet to be realized.

3   Automated measures of welfare as part of precision farming

The automated measures of welfare that are now being developed can potentially provide farmers with valuable information that they can then use to make decisions about how to manage their flocks for better health, for example, whether to put down extra litter or to apply medication. Using such technology, farmers are still in control but are given an extension of their own expertise in the form of extra eyes and ears that operate continuously when they cannot be present.

More ambitiously, however, the same information can also potentially be used to feed into more integrated automated control systems in which management decisions are taken largely or entirely by a computer. Such precision livestock systems (Werner et al., 2003; Banhazi and Black, 2009; Berckmans, 2017) have been inspired by the now-widespread use of precision agriculture in the growing of crops, where many different variables about both the environment and the growing plants are measured and then actions are automatically taken to match practice to crop needs with the goal of optimizing returns while preserving resources. Here, human decisions often defer to machine algorithms.

However, the idea that the concepts used to grow plants successfully can be transferred directly to poultry farming is controversial. On the one hand, Werkheiser (2018) argues that precision farming allows 'modern, large-scale farms to replicate and even improve on the benefits of caring farmers who know their animals, transferred to a much larger scale. This could be done via closer monitoring than farmers can provide to even a few animals, as well as integration of the data via decision algorithms that improve on the guesswork of traditional stockpersons'. On the other hand, precision farming may make the continued use of intensive systems more likely through making them more efficient and, therefore, more viable, which is of concern to some animal welfare advocates (Stevenson, 2017). For poultry, housed together in large groups and highly sensitive to the details of diet, temperature and air and litter quality, there is a vast potential for not only improving welfare through the use of technology but also reducing welfare through an emphasis on efficiency at the expense of welfare. While improving production and improving welfare are not mutually exclusive (Dawkins, 2016), they do not always coincide. When there is a conflict, what matters is the priority given to welfare.

There are encouraging signs that improving bird welfare is now a priority in much of the research to developing PLF for poultry. In a systematic review Rowe et al. (2019) concluded that 64% of PLF publications had animal health and welfare as one of their goals, the other being increased production. For the publications that only had one goal, more publications had animal health and welfare as the only goal of the study (40%) compared to production (27%). However, with PLF receiving increasing attention around the world and particularly in the USA, China and Belgium, there is an on-going need to ensure that animal welfare is seen as a key priority for sustainability in general and poultry production in particular (Rowe et al., 2019).

4 Why isn't automated welfare assessment more widely used?

The development of increasingly sophisticated monitoring systems for poultry farming with their potential to improve both welfare and productivity raises the question of why there are not more automated welfare monitoring systems in place on commercial poultry farms. This cannot be explained simply by farmers being unwilling to invest in technology because, as pointed out at the beginning of this chapter, many poultry houses now have complex computer-controlled heating and ventilation systems. Poultry farmers are clearly not averse to technology if it works and gives them something they could not obtain otherwise, It is the next step – the widespread use of automated ways of assessing behaviour and welfare – that is so far not been adopted by poultry producers.

Over 10 years ago, Wathes et al. (2008) suggested that this lack of take-up might be because the equipment was expensive or unreliable or because it was untested under commercial farm conditions, but there has been plenty of time since then for these problems to have been resolved. Yet poultry farming still appears to lag behind other sectors of agriculture in its use of basic health monitoring. In the dairy sector, for example, it is common to use technology to obtain diagnostic data about a range of health- and performance-related criteria (Caja et al., 2016). Poultry farming, with its need to monitor the health and welfare of large numbers of animals, its tight economic margins and the need for transparency in the face of public concern over bird welfare would seem to be even more in need of any help that technology can bring.

A possible answer for the fact that automated welfare is not more widely adopted is that the benefits – economic, production or even welfare – of available automated monitoring systems just simply have not been sufficiently demonstrated for farmers to see the point of investing in them. Caja et al. (2016) pointed out that 'dairy farmers will pay for and use technologies that provide what is, to them, a straightforward answer to a straightforward question (should I inseminate cow x?) when they believe it will have positive economic impact'. Although many of the poultry monitoring systems described previously have the potential to improve bird health and welfare, their actual benefits, over and above those that could be obtained by simpler methods, have not yet been demonstrated in practice. For example, keeping growing chicks within the correct temperature and humidity range is critical for their health and welfare (Dawkins et al., 2004), so it may be seen as much simpler to measure this directly with temperature and humidity sensors rather than indirectly by using the spatial distribution of the birds from cameras to determine whether they are too hot or too cold. In other words, good control of the environmental input may be more effective at maintaining the optimal temperature than measuring the results of temperature deviations using a complex and expensive camera or a sound system. If the aim is to monitor how much food chickens are eating, the most obvious way is to measure this directly rather than using the sound the birds make by pecking at food dishes. Water use, another variable that is easily monitored, has been recommended as an easy-to-use measure of welfare (RSPCA, 2013; OIE, 2016) since both increases and decreases in water consumption can indicate health problems and the total amount of water consumed by a flock over its lifetime is positively correlated with the prevalence of foot pad dermatitis at slaughter (Manning et al., 2007). Automated assessment of flock behaviour using optical flow also predicts hockburn and pododermatitis and is able to do so more effectively and when the birds are only a few days old (Dawkins et al., 2017) but for farmers to move from the relatively simple method of metering water to automated welfare monitoring with optical flow patterns, they would need to be convinced that the automated system gave them substantially better flock outcomes. Unless and until it can be demonstrated that there are additional benefits from automated welfare assessment, farmers will not invest in or use it. The current situation is that although there is a considerable potential for welfare improvement through automated monitoring, the evidence that welfare is actually improved by using such technology is still missing, even for the most basic health elements of welfare.

When it comes to more complex measures such as activity, or the recognition of specific behaviours such as dustbathing, there is even less evidence of benefit. Unless farmers can be convinced that knowing how much dustbathing or other behaviours their birds are actually doing helps them to keep their flocks healthier, or demonstrably improves the welfare of their birds in some tangible way, it is unlikely that they will invest in the automated ways of measuring it.

5 Conclusions

As this chapter shows, technology has a considerable potential for improving welfare through its ability to provide constant monitoring of the health and welfare of poultry throughout their lives and to provide early warning of problems before they become serious. The most promising existing systems are those that measure clearly defined health outcomes in broiler chickens such as lameness and leg and foot defects that are universally agreed to indicate poor welfare and also have economic implications for farmers, but even these have yet to be implemented on a commercial scale. Comparable systems for welfare monitoring in other poultry have yet to be developed.

6 Future trends

Looking into the future, prime targets for using automated monitoring include occasional events that can have serious welfare consequences such as smothering and feather-pecking in laying hens. 'Smothering' or 'piling'

(Campbell et al., 2016) is when hens sometimes pile on top of each other to the point that some of them may be killed, while feather-pecking is where hens damage the feathers of other birds, sometimes escalating to the point of injury or even death (Savory, 1995). The causes of these serious welfare issues are incompletely understood (Green et al., 2000; Gebhardt-Henrich and Stratmann, 2016) and although the risk of feather-pecking can be reduced by good management (Nicol et al., 2013), it still occurs, often with devastating effects. If automated monitoring of behaviour could give advance warning of when outbreaks are likely to occur or at least which flocks were most at risk (Lee et al., 2011), this would be a major contribution to improving poultry welfare.

Another use of automated systems could be to monitor the numbers of layers or broilers going outside to fulfil legal or auditing standard requirements for being 'free range' and to monitor a wider range of bird behaviours (Valletta et al., 2017) such as dustbathing and perching to satisfy a public demand for a fully comprehensive picture of bird welfare.

Properly used, automated welfare assessment could combine the wisdom of good stockmanship with far greater attention to individual birds than is currently possible. Smart computing and sensing technology offers the opportunity for shifting poultry production from an entirely group-based approach to a greater consideration of the needs of individual birds. While this shift has been relatively easy for sows and dairy cows where tagging can facilitate individual diets and treatments (Halachmi and Guarino, 2016), future developments may extend this to large groups of untagged poultry. Even crop agriculture does not consider a whole field as a single homogenous unit but treats different parts of the same field differently depending on soil type, moisture levels or the health of the plants growing there. Precision farming for poultry offers the possibility of being similarly 'smart' in not having to treat a flock as a single homogenous whole but rather to be able to apply heat, ventilation or extra litter at least down to a small area of the house.

Whatever the future holds for automated welfare assessment, however, its full potential has yet to be realised in practice. We have argued in this chapter that one of the main reasons for this is that its benefits have not been demonstrated sufficiently to convince poultry farmers to adopt the technology on a large scale. Currently, it is not seen as adding much, if anything, over and above what a stockperson could do on his or her own or with equipment that is already available, such as ventilation control and even water meters. Automated welfare assessment is not yet seen as a worth-while practice, either as a commercial proposition (Jukan et al., 2017), for increasing efficiency or even as a way of achieving its primary purpose of making real improvements to bird welfare. The following are suggested as ways of ensuring that automated monitoring fulfils its potential for achieving higher standards of poultry welfare within the context of sustainable food production:

1 The quality of information provided by automated monitoring needs to be improved above its present level so that it is clearly and demonstrably offering farmers something that is better than anything they have available at present. This means that the data obtained from sensors such as cameras and microphones should not be viewed in isolation but rather be part of a fully integrated system that includes all other kinds of information known to impact bird welfare, such as temperature, light and humidity (Van Hertem et al., 2018). To fully achieve this, new and better systems will be needed for analysing and sharing data sets (Jukan et al., 2017) so that best practice can be identified and recommended with the confidence that it is based on the largest possible evidence base. The collection and analysis of such 'Big Data' sets will not only mean better understanding of the factors leading to improved welfare but also the environmental, economic and food safety as well as other advantages of improving welfare (Gocsik et al., 2016; Henningsen et al., 2018). Linking welfare to these other benefits is an important part of making sure animal welfare is given high priority in the decision-making of farmers, governments and the public (Garnett et al., 2013; Dawkins, 2016; Jukan et al., 2017).

2 Automated welfare assessment needs to develop smarter algorithms so that it is both better at detecting welfare problems and able to include more in the definition of 'good welfare' than just the physical health outcomes it currently measures. There is a need for more powerful algorithms that take account of a wider range of welfare measure, but these will need to go hand in hand with more tightly defined and clearly demonstrated definitions of what good welfare is. An algorithm that correctly detects a specific behaviour will not carry much weight as a welfare indicator if that behaviour itself is not linked firmly to what is meant by good welfare, for example by being either an indicator of good health or something that the animals have been shown to want. In other words, the correct *interpretation* of what automated systems can tell us in welfare terms will be as important to their success as the development of new algorithms themselves.

Another key factor in the welfare value of automation will be the extent to which the technology is able to provide information about which interventions are most appropriate at any given time. Knowledgable systems that can tell farmers when there is a genuine problem with a flock and do so reliably without setting off too many false alarms will help the technology to be widely accepted. But even smarter systems that not only detect problems but can also indicate what actions are needed will be an important future development.

3   Public doubts about the welfare implications of automated assessment need to be taken seriously and directly addressed. This is best done by making sure that animal welfare achieves universally high priority with all stakeholders, which in turn means turning the potential welfare advantages of automated assessment into real improvements in the ways listed earlier. As has been emphasized throughout this chapter, automated welfare assessment offers a great promise of improving welfare through constant monitoring, greater care for the individual and early warning of problems and, through the use of large data sets, identifying best practice from around the world. Indeed, as automated measurements become more widespread and provide ever larger data sets, machine learning may provide us with new measures of 'welfare' itself based on a far wider range of measures than is currently possible and using information collected across animals' entire lifetimes.

In addition to improving bird welfare, automated systems have the potential to bring other benefits such as reduced susceptibility to diseases through improved immunity, earlier detection of infection, less waste through reduced mortality, higher standards of food quality and safety and greater job satisfaction for those in charge of poultry (Dawkins, 2016). But the promise has yet to be convincingly fulfilled. Changes in poultry production that make use of automated measurements are likely to continue in the face of rising demand for poultry products, increasing scarcity of human labour to look after birds and increasing public demands for higher welfare standards and transparency by producers.

It is important that the benefits that smart technology appears to offer us are ultimately translated into a reality of better lives for the birds themselves.

7   References

Aydin, A. 2017a. Development of an early detection system for lameness of broilers using computer vision. *Computers and Electronics in Agriculture* 136, 140-6. doi:10.1016/j.compag.2017.02.019.

Aydin, A. 2017b. Using 3D vision camera system to automatically assess the level of inactivity in broiler chickens. *Computers and Electronics in Agriculture* 135, 4-10. doi:10.1016/j.compag.2017.01.024.

Aydin, A. and Berckmans, D. 2016. Using sound technology to automatically detect the short-term feeding behaviours of broiler chickens. *Computers and Electronics in Agriculture* 121, 25-31. doi:10.1016/j.compag.2015.11.010.

Aydin, A., Bahr, C. and Berckmans, D. 2015. Automatic classification of measures of lying to assess the lameness of broilers. *Animal Welfare* 24(3), 335-43. doi:10.7120/09627286.24.3.335.

Banhazi, T. M. and Black, J. L. 2009. Precision livestock farming: a suite of electronic systems to ensure the application of best practice management on livestock farms. *Australian Journal of Multi-Disciplinary Engineering* 7(1), 1-14. doi:10.1080/144883 88.2009.11464794.

Banhazi, T. M., Lehr, H., Black, J. L., Crabtree, H., Schofield, P., Tscharke, M. and Berckmans, D. 2012. Precision Livestock Farming: an international review of scientific and commercial aspects. *International Journal of Agricultural and Biological Engineering* 5, 1-9.

Ben Sassi, N., Averos, X. and Estevez, I. 2016. Technology and poultry welfare. *Animals* 6(10), 62. doi:10.3390/ani6100062.

Berckmans, D. 2017. General introduction to precision livestock farming. *Animal Frontiers* 7(1), 6-11. doi:10.2527/af.2017.0102.

Bono, G. and Mori, B. D. 2005. Animals and their quality of life: considerations beyond mere welfare. *Veterinary Research Communications* 29(Suppl. 2), 165-8.

Brambell, F. W. R. 1965. Report of the technical committee to enquire into the welfare of animals kept under intensive husbandry systems. Doc. 2836. Her Majesty's Stationery Office, London, UK. Reprinted 1970.

Broom, D. M. 2007. Quality of Life means welfare: how is it related to other concepts and assessed? *Animal Welfare* 16, 43-53.

Buscher, W. 2019. Digitization of the barn. *Zuchtungskunde* 91, 35-44.

Caja, G., Castro-Costa, A. and Knight, C. H. 2016. Engineering to support wellbeing of dairy animals: background and current scenario. *Journal of Dairy Research* 83(2), 136-47. doi:10.1017/S0022029916000261.

Campbell, D. L. M., Makagon, M. M., Swanson, J. C. and Siegford, J. M. 2016. Litter use by laying hens in a commercial aviary: dustbathing and piling. *Poultry Science* 95(1), 164-75. doi:10.3382/ps/pev183.

Campbell, D. L. M., Hinch, G. N., Dyall, T. R., Warin, L., Little, B. A. and Lee, C. 2017. Outdoor stocking density in free-range laying hens: radio-frequency identification of impacts on range use. *Animal* 11(1), 121-30. doi:10.1017/S1751731116001154.

Carvalho, T. M. R., Massari, J. M., Sabino, L. A. and Moura, D. J. 2013. Sensor placement to reach thermal comfort and air quality in broiler housing. *Precision Livestock Farming 2013 - Papers Presented at the 6th European Conference on Precision Livestock Farming, ECPLF 2013*, pp. 945-52.

Chen, Y., Ni, J. Q., Diehl, C. A., Heber, A. J., Bogan, B. W. and Chai, L. L. 2010. Large scale application of vibration sensors for fan monitoring at commercial layer hen houses. *Sensors (Basel, Switzerland)* 10(12), 11590-604. doi:10.3390/s101211590.

Coelho, D. J. de R., Tinôco, I. de F. F., Baptista, F. J., Souza, C. de F., Sousa, F. C. de, Cruz, V. F. da, Vieira, M. de F. A., Mendes, M. A. dos S. A., Oliveira, K. P. de and Barbari, M. 2016. Mapping the thermal comfort index in laying hens facilities. Organising Committee, CIGR 2016, Aarhus, Denmark.

Colles, F. M., Cain, R. J., Nickson, T., Smith, A. L., Roberts, S. J., Maiden, M. C. J., Lunn, D. and Dawkins, M. S. 2016. Monitoring chick flock behaviour provides early warning of infection by human pathogen Campylobacter. *Proceedings of the Royal Society of London, Series B* 283(1822).

Cook, N. J., Smykot, A. B., Holm, D. E., Fasenko, G. and Church, J. S. 2006. Assessing feather cover of laying hens by infrared thermography. *Journal of Applied Poultry Research* 15(2), 274-9. doi:10.1093/japr/15.2.274.

Corkery, G., Ward, S., Kenny, C. and Hemmingway, P. 2013. Incorporating smart sensing technology into the poultry industry. *Journal of World's Poultry Research* 3, 106-28.

Curi, T. M. RdC., Conti, D., Vercellino, RdA., Massari, J. M., Moura, D. Jd, Souza, Z. Md and Montanari, R. 2017. Positioning of sensors for control of ventilation systems in broiler houses: a case study. *Scientia Agricola* 74(2), 101-9. doi:10.1590/1678-992x-2015-0369.

D'Alfonso, T. H., Manbeck, H. B. and Roush, W. B. 1996. A case study of temperature uniformity in three laying hen production buildings. *Transactions of the American Society of Agricultural Engineers* 3, 669.

Dawkins, M. S. 2016. Animal welfare and efficient farming: is conflict inevitable? *Animal Production Science* 57(2), 201. doi:10.1071/AN15383.

Dawkins, M. S., Donnelly, C. A. and Jones, T. A. 2004. Chicken welfare is influenced more by housing conditions than by stocking density. *Nature* 427(6972), 342-4. doi:10.1038/nature02226.

Dawkins, M. S., Lee, H.-J.,Waitt, C. D. and Roberts, S. J. 2009. Optical flow patterns in broiler chicken flocks as automated measures of behaviour and gait. *Applied Animal Behaviour Science* 119(3-4), 203-9. doi:10.1016/j.applanim.2009.04.009.

Dawkins, M. S., Cain, R. and Roberts, S. J. 2012. Optical flow, flock behaviour and chicken welfare. *Animal Behaviour* 84(1), 219-23. doi:10.1016/j.anbehav.2012.04.036.

Dawkins, M. S., Roberts, S. J., Cain, R. J., Nickson, T. and Donnelly, C. A. 2017. Early warning of footpad dermatitis and hock burn in broiler chicken flocks using optical flow, body weight and water consumption. *Veterinary Record* 180(20), 499. doi:10.1136/vr.104066.

de Jong, I. C.,van Harn, J., Gunnink, H., Hindle, V. A. and Lourens, A. 2012. Footpad dermatitis in Dutch broiler flocks: prevalence and factors of influence. *Poultry Science* 91(7), 1569-74. doi:10.3382/ps.2012-02156.

De Montis, A., Pinna, A., Barra, M. and Vranken, E. 2013. Analysis of poultry eating and drinking behavior by software eYeNamic. *Journal of Agricultural Engineering* 44(2s), 166-72. doi:10.4081/jae.2013.275.

Drake, K. A.,Donnelly, C. A. and Dawkins, M. S. 2010. Influence of rearing and lay risk factors on propensity for feather damage in laying hens. *British Poultry Science* 51(6), 725-33. doi:10.1080/00071668.2010.528751.

Du, X., Lao, F. and Teng, G. 2018. A sound source localization analytical method for monitoring the abnormal night vocalisations of poultry. *Sensors* 18(9), Article No. 2906. doi:10.3390/s18092906.

Edgar, J. L., Nicol, C. J., Pugh, C. A. and Paul, E. S. 2013. Surface temperature changes in response to handling in domestic chickens. *Physiology and Behavior* 119, 195-200. doi:10.1016/j.physbeh.2013.06.020.

Ekstrand, C., Carpenter, T. E., Andersson, I. and Algers, B. 1998. Prevalence and control of foot-pad dermatitis in broilers in Sweden. *British Poultry Science* 39(3), 318-24. doi:10.1080/00071669888845.

Fernandez, A. P., Norton, T., Tullo, E., van Hertem, T., Youssef, A., Exadaktylos, V., Vranken, E., Guarino, M. and Berckmans, D. 2018. Real-time monitoring of broiler flock's welfare status using camera-based technology. *Biosystems Engineering* 173, 103-14. doi:10.1016/j.biosystemseng.2018.05.008.

Fleet, D. J. and Weiss, Y. 2005. Optimal flow estimation. In: Paragios, N., Chen, Y. and Faugeras, O. (Eds), *Mathematical Models for Computer Vision*. Springer, New York, NY, pp. 239-58.

Fournel, S., Rousseau, A. N. and Laberge, B. 2017. Rethinking environmental control strategy of confined animal housing systems through precision livestock farming. *Biosystems Engineering* 155, 96–123. doi:10.1016/j.biosystemseng.2016.12.005.

Fraser, D. 2008. *Understanding Animal Welfare: the Science in Its Cultural Context*. Wiley-Blackwell, Oxford, UK.

Fraser, D., Duncan, I. J. H., Edwards, S. A., Grandon, T., Gregory, N. G., Guyonnet, V., Hemsworth, P. H., Huertas, S. M., Huzzey, J. M., Mellor, D. J., Mench, J. A., Spinka, M. and Whay, H. R. 2013. General principles for the welfare of animals in production systems: the underlying science and its application. *Veterinary Journal* 198(1), 19–27. doi:10.1016/j.tvjl.2013.06.028.

Garnett, T., Appleby, M. C., Balmford, A., Bateman, I. J., Benton, T. G., Bloomer, P., Burlingame, B., Dawkins, M., Dolan, L., Fraser, D., Herrero, M., Hoffmann, I., Smith, P., Thornton, P. K., Toulmin, C., Vermeulen, S. J. and Godfray, H. C. 2013. Sustainable intensification in agriculture: premises and policies. *Science* 341(6141), 33–4. doi:10.1126/science.1234485.

Gebhardt-Henrich, S. G. and Stratmann, A. 2016. What is causing smothering in laying hens? *Veterinary Record* 179(10), 250–1. doi:10.1136/vr.i4618.

Gocsik, É., Brooshooft, S. D., de Jong, I. C. and Saatkamp, H. W. 2016. Cost-efficiency of animal welfare in broiler production: a pilot study is the Welfare Quality® assessment protocol. *Agricultural Systems* 146, 55–69. doi:10.1016/j.agsy.2016.04.001.

Green, L. E., Lewis, K., Kimpton, A. and Nicol, C. J. 2000. Cross-sectional study of the prevalence of feather pecking of laying hens in alternative systems and its associations with management and disease. *Veterinary Record* 147(9), 233–8. doi:10.1136/vr.147.9.233.

Gunnarsson, S., Keeling, L. J. and Svedberg, J. 1999. Effect of rearing factors on the prevalence of floor eggs, cloacal cannibalism and feather pecking in commercial flocks of loose housed laying hens. *British Poultry Science* 40(1), 12–8. doi:10.1080/00071669987773.

Halachmi, I. and Guarino, M. 2016. Editorial: precision livestock farming: a 'per animal' approach using advanced monitoring technologies. *Animal* 10(9), 1482–3. doi:10.1017/S1751731116001142.

Henningsen, A., Czekaj, T. G., Forkman, B., Lund, M. and Nielsen, A. S. 2018. The relationship between animal welfare and economic perfomance at farm level: a quantitative study of Danish pig producers. *Journal of Agricultural Economics* 69(1), 142–62. doi:10.1111/1477-9552.12228.

Ji, B., Zheng, W., Gates, R. S. and Green, A. R. 2016. Design and performance evaluation of the upgraded portable monitoring unit for air quality in animal housing. *Computers and Electronics in Agriculture* 124, 132–40. doi:10.1016/j.compag.2016.03.030.

Jones, T. A., Donnelly, C. A. and Dawkins, M. S. 2005. Environmental and management factors affecting the welfare of chickens on commercial farms in the United Kingdom and Denmark stocked at five densities. *Poultry Science* 84(8), 1155–65. doi:10.1093/ps/84.8.1155.

Jukan, A., Masip-Bruin, X. and Amla, N. 2017. Smart computing and sensing technologies for animal welfare: a systematic review. *ACM Computing Surveys* 50(1), 1–27. doi:10.1145/3041960.

Kashiha, M. A., Green, A. R., Sales, T. G., Bahr, C., Berckmans, D. and Gates, R. S. 2015. Application of image processing on hen tracking in an environmental preference chamber. *Precision Livestock Farming 2015 - Papers Presented at the 7th European Conference on Precision Livestock Farming*, ECPLF 2015, pp. 185-94.

Kestin, S. C., Knowles, T. G., Tinch, A. E. and Gregory, N. G. 1992. Prevalence of leg weakness in broiler chickens and the relationship with genotype. *Veterinary Record* 133, 190-4.

Lee, H. J., Roberts, S. J., Drake, K . A. and Dawkins, M. S. 2011. Prediction of feather damage in laying hens using optical flows and Markov models. *Journal of the Royal Society, Interface* 8(57), 489-99. doi:10.1098/rsif.2010.0268.

Lee, J., Noh, B., Jang, S., Park, D., Chung, Y. and Chang, H. H. 2015. Stress detection and classification of laying hens by sound analysis. *Asian-Australasian Journal of Animal Sciences* 28(4), 592-8. doi:10.5713/ajas.14.0654.

Leroy, T., Vranken, E., Van Brecht, A., Struelens, E., Sonck, B. and Berckmans, D. 2006. A computer vision method for on-line behavioral quantification of individually caged poultry. *Transactions of the ASABE* 49(3), 795-802. doi:10.13031/2013.20462.

Liakos, K. G., Busato, P., Moshou, D., Pearson, S. and Bochtis, D. 2018. Machine learning in agriculture: a review. *Sensors* 18(8), Article No. 2674. doi:10.3390/s18082674.

Main, D. C. J., Mullan, S., Atkinson, C., Cooper, M., Wrathhall, J. H. M. and Blokhuis, H. J. 2014. Best practice framework for animal welfare certification schemes. *Trends in Food Science and Technology* 37(2), 127-36. doi:10.1016/j.tifs.2014.03.009.

Manning, L., Chadd, S. A. and Baines, R. N. 2007. Water consumption in broiler chickens: a welfare indicator. *World's Poultry Science Journal* 63(1), 63-71. doi:10.1017/S0043933907001274.

Mason, G. and Mendl, M. 1993. Why is there no simple way of measuring animal welfare? *Animal Welfare* 2, 301-20.

McCafferty, D. J. 2013. Applications of thermal imaging in avian science. *Ibis* 155(1), 4-15. doi:10.1111/ibi.12010.

Mellor, D. J. 2016. Updating animal welfare thinking: moving beyond the 'Five Freedoms' towards 'a life worth living'. *Animals* 6(3), 21. doi:10.3390/ani6030021.

Mellor, D. J. and Beausoleil, N. J. 2015. Extending the 'Five Domains' model for animal welfare assessment to incorporate positive welfare states. *Animal Welfare* 24(3), 241-53. doi:10.7120/09627286.24.3.241.

Moe, R. O., Bohlin, J., Flo, A., Vasdal, G. and Stubsjoen, S. M. 2017. Hot chicks, cold feet. *Physiology and Behavior* 179, 42-8. doi:10.1016/j.physbeh.2017.05.025.

Morota, G., Ventura, R. V., Silva, F. F., Koyama, M. and Fernando, S. C. 2018. Machine learning and data mining advance predictive big data in precision animal agriculture. *Journal of Animal Science* 96(4), 1540-50. doi:10.1093/jas/sky014.

Moura, C. J., Naas,I., Alves, E. C. S., Carvalho, T. M. R., Vale, M. M. and Lima, K. A. O. 2008. *Scientific Agriculture* 119, 178-83.

Nasirahmadi, A., Edwards, S. A. and Sturm, B. 2017. Implementation of machine vision for detecting behaviour of cattle and pigs. *Livestock Science* 202, 25-38. doi:10.1016/j.livsci.2017.05.014.

Nicol, C. J., Bestman, M.,Gilani, A. M., De Haas, E. N., De Jong, I. C., Lambton, S., Wagenaar, J. P., Weeks, C. A. and Rodenburg, T. B. 2013. The prevention and control of feather pecking: applications to commercial systems. *World's Poultry Science Journal* 69(4), 775-88. doi:10.1017/S0043933913000809.

Norton, T., Vranken, E., Exadaktylos, V., Berckmans, D., Lehr, H., Vessier, I., Blokhuis, H. and Berckmans, D. 2016. Implementation of Precision Livestock Farming (PLF) technology on EU farms: results from the EU-PLF project. In: *CIGR-AgEng Conference*, 26-29 June 2016, Aarhus, Denmark. Abstracts and full papers, 2016, pp. 1-7.

Pereira, D. F., Nääs, I. D. A., Gabriel Filho, L. R. A. and Neto, M. M. 2012. Cluster index for accessing thermal comfort for broiler breeders. *ASABE - 9th International Livestock Environment Symposium 2012*, ILES 2012, pp. 207-12.

Pereira, D. F., Miyamoto, B. C. B., Maia, G. D. N., Tatiana Sales, G., Magalhães, M. M. and Gates, R. S. 2013. Machine vision to identify broiler breeder behavior. *Computers and Electronics in Agriculture* 99, 194-9. doi:10.1016/j.compag.2013.09.012.

Pereira, E. M., Naas, IdA. and Garcia, R. G. 2014. Identification of acoustic parameters for broiler welfare estimates. *Engenharia Agricola* 34(3), 413-21. doi:10.1590/S0100-69162014000300004.

OIE. 2016. Chapter 7. Animal Welfare and broiler chicken production systems. Terrestrial Animal Health Code. OIE. World Organisation for Animal Health, pp. 1-7. Available at: http://www.oie.int/index.php?id=169&L=0&htmfile=chapitre_aw:broiler_chicken.ht (accessed on 08 July 2016).

Roberts, S. J., Cain, R. and Dawkins, M. S. 2012. Prediction of welfare outcomes for broiler chickens using Bayesian regression on continuous optical flow data. *Journal of the Royal Society, Interface* 9(77), 3436-43. doi:10.1098/rsif.2012.0594.

Rowe, E., Dawkins, M. S. and Gebhardt-Heinrich, S. 2019. A systematic review of precision livestock farming in the poultry sector: is technology focussed on improving animal welfare? *Animals* 9(9), Article 614.

RSPCA. 2013. *Welfare Standards for Chickens*. Royal Society for the Prevention of Cruelty to Animals, Horsham, UK.

Savory, C. J. 1995. Feather pecking and cannibalism. *World's Poultry Science Journal* 51(2), 215-19. doi:10.1079/WPS19950016.

Scott, E. M., Nolan, A., Reid, J. and Wiseman-Orr, M. L. 2007. Can we really measure animal QoL? Methodologies for measuring QoI in people and other animals. *Animal Welfare* 16, 17-24.

Sonka, M., Hlavac, V. and Boyle, R. 1999. *Image Processing Analysis and Machine Vision* (2nd edn.). PWS Publishing, London, UK.

Stadig, L. M., Ampe, B., Rodenburg, T. B., Reubens, B., Maselyne, J., Zhuang, S., Criel, J. and Tuyttens, F. A. M. 2018. An automated positioning system for monitoring chickens' location: accuracy and registration success in a free-range area. *Applied Animal Behaviour Science* 201, 31-9. doi:10.1016/j.applanim.2017.12.010.

Stevenson, P. 2017. Precision livestock farming: could it drive the livestock sector in the wrong direction? Available at: https://www.ciwf.org.uk/media/7431928/plf-could-it-drive-the-livestock-sector-in-the-wrong-direction.pdf (accessed on 9 May 2019).

Taylor, K. D. and Mills, D. S. 2007. Is quality of life a useful concept for companion animals? *Animal Welfare* 16, 55-65.

Valletta, J. J., Torney, C., Kings, M., Throbton, A. and Madden, J. 2017. Applications of Machine learning in animal behaviour studies. *Animal Behaviour* 124, 203-20. doi:10.1016/j.anbehav.2016.12.005.

Vanderhasselt, R. F., Sprenger, M., Duchateau, L. and Tuyttens, F. A. 2013. Automated assessment of footpad dermatitis in broiler chickens at the slaughterline: evaluation and correspondence with human expert scores. *Poultry Science* 92(1), 12-8. doi:10.3382/ps.2012-02153.

Van Hertem, T., Rooijahkers, L., Berckmans, D., Peña Fernández, A., Norton, T., Berckmans, D. and Vranken, E. 2017. Appropriate data visualisation is key to Precision Livestock

Farming acceptance. *Computers and Electronics in Agriculture* 138, 1-10. doi:10.1016/j.compag.2017.04.003.

Van Hertem, T., Norton, T., Berckmans, D. and Vranken, E. 2018. Predicting broiler gait scores from activity monitoring and flock data. *Biosystems Engineering* 173, 93-102. doi:10.1016/j.biosystemseng.2018.07.002.

Wathes, C. M. 2010. Lives worth living? *Veterinary Record* 166(15), 468-9. doi:10.1136/vr.c849.

Wathes, C. M., Kristensen, H. H., Aerts, J.-M. and Berckmans, D. 2008. Is precision livestock farming an engineer's daydream or nightmare, an animal's friend of foe and a farmer's panacea or pitfall? *Computers and Electronics in Agriculture* 64(1), 2-10. doi:10.1016/j.compag.2008.05.005.

Webster, A. J. F. 2001. Farm animal welfare: the five freedoms and the free market. *Veterinary Journal* 161(3), 229-37. doi:10.1053/tvjl.2000.0563.

Welfare Quality. 2009. *Assessment Protocol for Poultry, Broiler and Laying Hens*. Welfare Quality, Lelystad, The Netherlands. Available at: http://www.animalwelfareplatfo rm.eu/Twelve-farm-animal-welfare-criteria.php.

Werkheiser, I. 2018. Precision livestock farming and farmers' duties to livestock. *Journal of Agricultural and Environmental Ethics* 31(2), 181-95. doi:10.1007/s10806-018-9720-0.

Werner, A., Jarfe, A., Stafford, J. V. and Cox, S. W. R. 2003. *European Conference on Precision Livestock Farming (1st : 2003 : Berlin G) and European Conference on Precision Agriculture (4th : 2003 : Berlin G) 2003 Programme Book of the Joint Conference of ECPA-ECPLF*. Wageningen Academic, Wageningen, the Netherlands.

Winckler, C. 2019. Assessing animal welfare at the farm level: do we care sufficiently about the individual? *Animal Welfare* 28(1), 77-82. doi:10.7120/09627286.28.1.077.

Wood-Gush, D. G. M. 1971. *The Behaviour of the Domestic Fowl*. Heinemann, London, UK.

Yahav, S., Shinder, D., Tanny, J. and Cohen, S. 2005. Sensible heat loss: the broiler's paradox. *World's Poultry Science Journal* 61(3), 419-34. doi:10.1079/WPS200453.

Zaninelli, M., Redaelli,V., Tirloni, E., Bernardi, C., Dell'Orto, V. and Savoini, G. 2016. First results of a detection sensor for the monitoring of laying hens reared in a commercial organic egg production farm based on the use of infrared technology. *Sensors (Basel, Switzerland)* 16(10). doi:10.3390/s16101757.

Zaninelli, M., Redaelli, V., Luzi, F., Bontempo, V., Dell'Orto, V. and Savoini, G. 2017. A monitoring system for laying hens that uses a detection sensor based on infrared technology and image pattern recognition. *Sensors* 17(6). doi:10.3390/s17061195.

Zaninelli, M., Redaelli, V., Luzi, F., Mitchell, M., Bontempo, V., Cattaneo, D., Dell'Orto, V. and Savoini, G. 2018. Development of a machine vision method for the monitoring of laying hens and detection of multiple nest occupations. *Sensors* 18(1). doi:10.3390/s18010132.

Zhao, Y., Aarnink, A. J. A., Hofschreuder, P. and Groot Koerkamp, P. W. G. 2009. Evaluation of an impaction and a cyclone pre-separator for sampling high PM10 and PM2.5 concentrations in livestock houses. *Journal of Aerosol Science* 40, 868-78.

Zhuang, X. L., Bi, M. N., Guo,J. L.,Wu, S. Y. and Zhang, T. 2018. Development of an early warning algorithm to detect sick broilers. *Computers and Electronics in Agriculture* 144, 102-13. doi:10.1016/j.compag.2017.11.032.

Zimmerman, P. H., Koene, P. and Van Hooff, J. A. 2000. The vocal expression of feeding motivation and frustration in domestic laying hens. *Applied Animal Behaviour Science* 69, 265-73. doi:10.1016/s0168-1591(00)00136-2.